The Dynamics of Iranian Borders

Mansoureh Ebrahimi · Masoumeh Rad Goudarzi
Kamaruzaman Yusoff
Editors

The Dynamics of Iranian Borders

Issues of Contention

Springer

Editors
Mansoureh Ebrahimi
Faculty of Islamic Civilization
Universiti Teknologi Malaysia
Skudai, Johor
Malaysia

Kamaruzaman Yusoff
Faculty of Islamic Civilization
Universiti Teknologi Malaysia
Skudai, Johor
Malaysia

Masoumeh Rad Goudarzi
Department of Political Science,
Faculty of Literature and Humanities
University of Guilan
Rasht, Guilan
Iran

ISBN 978-3-030-07871-3 ISBN 978-3-319-89836-0 (eBook)
https://doi.org/10.1007/978-3-319-89836-0

Copyediting: PD Dr. Hans Günter Brauch, AFES-PRESS e.V., Mosbach, Germany
English Language Editor: Vanessa Greatorex, England

Printed on acid-free paper

This Springer imprint is published by the registered company Springer International Publishing AG part of Springer Nature
The registered company address is: Gewerbestrasse 11, 6330 Cham, Switzerland

Preface

Cross-border conflicts are common and generally known the world over, but those involving the traditional territories of Iran have received little attention. These include the Caspian Sea's legal regime in the north; the sovereignty of both Great and Little Tunb and Abu Musa Islands in the Persian Gulf; the use of the Hirmand River's water in the east; and the bloody war with Iraq (1980–1988) over maritime boundaries in the Arvand River. Over the last two hundred-plus years, Iran has lost vast territories and, time and again, has been forced to accept imposed borders that embrace ethnic and religious divisions that continue to this day. The authors therefore focus on a little-known historicity that fashioned current Iranian borders and also offer detailed assessments of the significant regional and international dynamics at play, including ongoing disputes and influences most likely to affect the future.

Necessity demands close attention to Iran's geopolitical position, which has always hosted conflicts between great powers and invited trans-regional sojourners in search of conquest and intrigues that attend the endless struggle for political and economic hegemony over rivals and locals for natural resources and sundry military strategies. The importance of Iran's position for marine transport between Asia and Europe has attracted the venality of colonial and regional greed with consequent political–military tensions for more than five hundred years. Hence, the region's enormous oil and gas reserves located between the Caucasus and Persian Gulf arrived a bit late in the game to prompt drawn-out contention over ports in the Strait of Hormuz and the Sea of Oman. These factors have not abandoned their influence on Iran's current borders and will certainly play major roles in the future.

The quality and stability of Iranian government(s) have also been subject to political and territorial integrity. The Safavid Empire provided a commanding central government with military stability that preserved traditional sovereignty and even restored various parts of the country. However, since that era, various political, military and economic frailties encouraged ethno-religious separatism, riots and a stream of crises, all of which central governments failed to effectively address. Territorial sovereignty also suffered insult, largely due to an inadequate leadership that was shorn of understanding for regional and trans-regional powers, especially Russia, Britain and France. Moreover, over time, Iran's governors, ministers and others came to depend on one power or the other, which incessantly facilitated diverse territorial losses. These losses include the three South Caucasus countries of Azerbaijan, Armenia and Georgia, as well as Afghanistan, Iraq and, most recently, Bahrain. Iran's acceptance of these forfeitures as the result of internal and external actors discomfits an ancient legacy.

From 1555 to 1975, eighteen contracts were signed between Iran and the Ottoman Empire and Iraq over its western borders. Treaty numbers and the length of this dispute reflect the nature of Iran's conflict with western neighbours, implying that involved parties were never satisfied and that grievances continued over various impositions, one after the other. But what is most significant in this regard is that this same period largely covers the establishment of European influence in the region, especially of Britain. Hence, the authors discuss these stimuli along with never-ending disputes, extensive wars and ethno-religious divisions that accompanied blatant conquest campaigns. Iran's weaknesses resulted in misdirected strength and naiveté as well as treachery and contributed much to a web of encroachment that also consumed traditional indigenous neighbours.

Thus, the entry of colonials into the Middle East and the Indian Ocean joined Czarist Russia's territorial expansion and opened a new chapter. British dominion in India brought numerous political changes with long-term military intrusions in the Sea of Oman and Persian Gulf, as well as the eastern regions of Iran that became Afghanistan. The UK sedulously prevented Iran from extending its power in the east to ensure the impossibility of any invasion of India by Czarist Russia. For this reason, colonial agents provoked Afghan tribal leaders against Iran and provided necessary pretexts and arms for independence. The consolidation of British power in India increased the economic importance of British control over the Persian Gulf and Sea of Oman as fundamental to the UK's hold on India. Consequently, this began a process of undermining Iran's influence in its southern regions, including the Persian Gulf and, in particular, the island of Bahrain.

Nevertheless, the discovery of oil justified a new thesis on the strategic importance of the Persian Gulf, thus converting the region from a transit hub to a capital source of energy for the economic and military benefit of the West. What distinguished this era from previous periods was Britain's massive military presence in the Persian Gulf and Sea of Oman, which also reflected Iran's decline and inability to consolidate and protect its western and southern borders. The 1937 Frontier Agreement and annexed Protocol finally allowed the UK to draw the border between Iran and Iraq in the Arvand River. Moreover, Britain's signing of military agreements with Bahrain's Sheikhs permitted their military presence on the island and virtually guaranteed British dominion over the region's oil and gas resources.

Similar extrinsic encroachments and internal factors shaped Iranian borders east, west and south. Iran's central government became anaemic as the country's resources and geopolitical importance allowed great powers to gnaw away at present and future cartographic speculation, including Iran's maritime boundaries in the Caspian Sea. Before the collapse of the Soviet Union in 1991, the Caspian Sea had been known as the 'Iranian-Soviet Sea' in accordance with the treaties between both countries (1921 and 1940), which considered the inland Sea a common maritime resource. In practice, the Soviet government was far more influential, especially since it surrounded the Sea on three sides. However, after the collapse of the Soviet Union, major geopolitical changes took place with the emergence of newly independent states as littoral neighbours. Attitudes and the legal regime thus became subject to new authorities. The Sea suddenly boasted five coastal nations, all laying claim to significant oil and gas resources; all holding different perspectives and aspirations; all determined to attempt governance, ownership and boundaries—and this is not to mention the greater interests of transnational powers.

This melee over the legal regime exemplifies the economic and geopolitical struggle for resources, and thus, it will have major impacts on trans-regional engagement in the Caspian Sea and the delineation of boundaries between littoral neighbours. The USA and the Federal Republic of Russia consistently pursue superior status to secure and increase their economic power and influence in the world's geopolitically strategic regions. The Caspian Sea is no exception, especially as it lies in the heart of Eurasia and the Caucasus, with massive hydrocarbon reserves. Thus, both countries do their utmost to influence the new legal regime in pursuit of respective interests. Nevertheless, indigenous littoral nations fail to understand and do not fully grasp the significance of their circumstances. Hence, any finalization and determination of Iran's share in the Caspian Basin depend on factors reaching far beyond regional agreements.

Following the Soviet collapse, Iranian authorities initially considered both the Caspian Basin and Caucasus of minor concern with respect to foreign policy. They mostly focused their attentions on the Middle East and North Africa. However, on discovery of the Caspian's vast energy reserves and attendant geopolitical importance, they paid special attention to regional developments to preserve Iran's economic and political interests, especially since participation in the new legal regime would heighten Iran's international and economic profiles. From then on, Iran deemed its competitive presence in the Caspian region 'vital to national interests', having direct effects on national security.

Global transformations after the fall of the USSR found the USA struggling to formulate a justifiable unipolar system of hegemony. America considers the Caspian Sea one of two major world energy reserves for the twenty-first century and attaches great importance to the region. Oil and geopolitics have granted premium standing to the Caspian Basin for diversified energy transfers to many who seek energy security. Thus, ideological and political differences between the USA and Iran will certainly affect final attributions and distributions of Caspian Sea borders and resources for Iran.

To conclude, Iran's unique geographical location has always been a cause of contention. The resulting crises attendant on its geopolitical importance represent a continuum of international competition for political gains and economic benefit. Iran's massive borders and evolving political weakness, along with influences from the Shahs of Qajar that maintained and strengthened Britain's hegemony in the region, were major factors affecting ongoing regional conflicts. Additional roles played by other world powers, such as France, Russia and the USA, are also noted.

Conflicts, unrest and regional wars were all consequences arising from power struggles that led to treaties and international agreements between Iran, Britain and Russia that caused the eventual loss of traditional Iranian territories. Hence, extrinsic impositions on Iran are the subject of this study as authors examine the turbulent climate that altered Iranian borders during the Qajar Dynasty.

The Dynamics of Iranian Borders: Issues of Contention provides political and historical analyses of Iran's geopolitical position in seven chapters that illustrate why the dominion of Persia became a key objective for Britain, Russia and the Ottomans. This volume presents an introduction, five essays and a conclusion on Iran's boundary disputes with its neighbours. It is a compilation of research which investigated Iran's nineteenth-century frontiers and loss of traditional territories. The authors elaborate the 'how' and 'why' these strategic borderlands became detached from the Iranian heartland over the last 200 years. The contributors present research findings that provide a wealth of hitherto little-known perspective.

Physiographic map of Iran (2001). *Source* This map is in the public domain. Courtesy of the Perry-Castañeda Library Map Collection, University of Texas Libraries, The University of Texas at Austin.

Political map of Iran (2001). *Source* This map is in the public domain. Courtesy of the Perry-Castañeda Library Map Collection, University of Texas Libraries, The University of Texas at Austin.

Johor Bahru, Johor, Malaysia
May 2018

Mansoureh Ebrahimi
Masoumeh Rad Goudarzi
Kamaruzaman Yusoff

Acknowledgements

This volume is the result of discussions between editors who observed that keeping national borders safe and secure was a sound strategy that avoided turmoil and even civil war. Hence, we undertook the necessary historical research with a focus on contentious border issues, especially regarding Iran's geopolitical and geostrategic location in the Middle East, which has been a battlefield for national and transnational interests over the centuries. The authors specifically highlight forgotten and/or marginalized sources in this volume with a view to helping to prevent future conflicts. The editors are incredibly grateful for the authors' prized contributions to the completion of this book.

Our heartfelt gratitude extends to the editorial expertise of PD Dr. Hans Günter Brauch, editor of *The Anthropocene: Politik—Economics—Society—Science (APESS)*. His dedication to excellence encouraged us with valuable insights that moved us forward and strengthened our respective commitments to finish this work. Our thanks also go to Dr. Johanna Schwarz, senior editor, Springer and Ms. Doerthe Mennecke-Buehler, Springer's producer in Heidelberg, as well as to the entire production team at Springer Nature in Chennai, Tamil Nadu, India. We are also indebted to *Springer Nature* and their publishing committee for their interest in this volume. We also offer sincere appreciation to Dr. Omar Zaid, MD, Sr. Lecturer (retired), Morals, Ethics and Values, UNITAR International University, KL, Malaysia, for painstakingly editing the initial draft and respectfully extend immense appreciation to Vanessa Greatorex (England) for a thoroughgoing final language edit.

With humble thanks, we send our deepest regard to three anonymous reviewers of the manuscript whose invaluable suggestions and comments significantly advanced our efforts. We also acknowledge Universiti Teknologi Malaysia Research Management Centre for the financial support (12H19 and 16H48 grants) that made this research possible. Finally, the authors assume full responsibility for inaccuracies, inadequacies, insufficiencies or views expressed herein.

Johor Bahru, Johor, Malaysia
May 2018

Contents

Editors and Contributors

About the Editors

Mansoureh Ebrahimi (Iran/Malaysia), Ph.D., is a senior lecturer with the Faculty of Islamic Civilization, Universiti Teknologi Malaysia. She has written about Iran's culture, history, religious beliefs and Malaysia's *halal* industry from the Islamic perspective. She also teaches Persian and Philosophy of Science and Civilization. Her books include *The British Role in Iranian Domestic Politics (1951–1953)* (Germany: Springer Nature, 2016); Mansoureh Ebrahimi and Kamaruzaman Yusoff (Eds.): *The Halal Industry in Malaysia: Prospects and Challenges* (Malaysia: UTM Press, 2015); she has also contributed several articles on culture, religion and history (including Malaysia, Middle East, Iran and Turkey) to reputable peer-reviewed journals; 'The British plot to remove Iran's democratically elected Prime Minister from power', in: *Acta Histriae* (Republic of Slovenia: Annales Publishing House), 23, (2015): 735–748.
Address: Mansoureh Ebrahimi, Ph.D., Faculty of Islamic Civilization, Universiti Teknologi Malaysia, 81310 Johor Bahru, Johor, Malaysia.
e-mail: mansoureh@utm.my and suriebrahimi@gmail.com *Website*: https://www.linkedin.com/in/suri-ebrahimi-08aa0457/; http://islamic.utm.my/mansoureh/ and http://afes-press-books.de/html/APESS_29.htm.

Masoumeh Rad Goudarzi (Iran) is assistant professor at the Department of Political Science, Faculty of Literature and Humanities, University of Guilan (Iran). She specialises in Islamic Political Thought and Politics and Government in the Middle East and Caucasus. She holds a BA in political Science from Shahid Beheshti University, Tehran; an MA from Tehran Islamic Azad University; and a Ph.D. in Politics and Government from UPM, Malaysia. She is the author of *Ayatollah Khomeini and the Subject of Islamic Government* (Germany: Lambert Academic Publishing, 2011); she has received several research awards and is published in reputable peer-reviewed journals.

Address: Department of Political Science, Faculty of Literature and Humanities, University of Guilan, Iran.
e-mail: Raad@guilan.ac.ir *Website*: http://staff.guilan.ac.ir/radghodarzi/?lg=0.

Kamaruzaman Yusoff (Malaysia) is professor at the Faculty of Islamic Civilization, Universiti Teknologi Malaysia, Head of West Asia Civilization Research Group, where he specialises in Middle Eastern History and Islamic Studies. He holds a BA in History from the National University of Malaysia; and an M.Litt. and a D.Litt. from the Department of Islamic and Middle Eastern Studies, Edinburgh University, Scotland. He is the author of several articles and books on the Middle East, Orientalism in Islamic History, and the relationship between the Ottoman Empire and the Malay World and has served as Head of Department Research Group and Dean of the Faculty.

Address: Kamaruzaman Yusoff, Ph.D., Faculty of Islamic Civilization, Universiti Teknologi Malaysia, 81310 Johor Bahru, Johor, Malaysia.
e-mail: y.kamaruzaman@utm.my and zuryati@yahoo.com.

Contributors

Abdollah Baei Lashaki Department of Law, Applied Art and Culture Center, Science—Education University, Nowshahr, Iran

Saeid Bahrami Department of Political Science, Faculty of Literature and Humanities, University of Guilan, Rasht, Iran

Mansoureh Ebrahimi Faculty of Islamic Civilization, Universiti Teknologi Malaysia, Johor Bahru, Johor, Malaysia

Mehrdad Fallah University of Mazandaran, Babolsar, Mazandaran, Iran

Saikou Kawsu Gassama Faculty of Islamic Civilization, Universiti Teknologi Malaysia, Johor Bahru, Johor, Malaysia

Dayyan Janbaaz Hadishahr Unit, Payam-e Nour University, Babolsar, Iran

Alireza Najafinejad Department of Political Science, Faculty of Humanities, Golestan University, Gorgan, Iran

Sediqeh Nazarpour University of Mazandaran, Babolsar, Mazandaran, Iran

Masoumeh Rad Goudarzi Department of Political Science, Faculty of Literature and Humanities, University of Guilan, Rasht, Guilan, Iran

Kamaruzaman Yusoff Faculty of Islamic Civilization, Universiti Teknologi Malaysia, Johor Bahru, Johor, Malaysia

Chapter 1
Introduction: The Russian Heartland Versus the British World Island

Mansoureh Ebrahimi, Saikou Kawsu Gassama and Kamaruzaman Yusoff

The themes discussed hold meaningful connections between geography, state power (economic, military and political) and political decision-making in the scramble between Russia and Britain to control and delineate Persian territory. The objective of our introduction is to examine political geography within the context of nineteenth-century colonial contests for power, influence, territory and resources. It primarily concerns Russia and Britain and their interests in Persia,[1] Central Asia and the Caspian Sea, with particular reference to Iranian borders. In the main, an attempt is made to apply Halford Mackinder's '*The Heartland* versus *The World-Island*' thesis as the struggle between Russia and Britain. However, other International Relations theories such as Classical Realism are also utilised to broaden our understanding of international competition for strategic resources from the eighteenth through to the twentieth century in these regions. Moreover, the intricate struggle between Britain and Russia over dominion of the lapsing Persian Empire requires that we revisit the concept of imperialism. For our purposes, Gerry Kearn's definition suffices:

> The use of resources of one place by people from another where the terms of extraction are set by the inequality in power between the two sets of peoples. This, then, would include both the displacement of local peoples by European colonialism (taking the land by force) as well as the exploitation of weak countries … through what is often termed unequal exchange (the resources are often taken on the cheap). (Kearns 1993: 9–10)

[1] 'Persia' is the previous name of Iran, which Reza Shah (Reza Khan, Riẓā Shāh) changed to 'Iran' in 1935. In this study, 'Iran' or 'Iranian Government' refers to all periods of time excepting certain treaties or specific phrases and quotation.

M. Ebrahimi (✉) · S. K. Gassama · K. Yusoff
Faculty of Islamic Civilization, Universiti Teknologi Malaysia, 81310 Johor Bahru, Johor, Malaysia
e-mail: mansoureh@utm.my; suriebrahimi@gmail.com
URL: https://www.linkedin.com/in/suri-ebrahimi-08aa0457/
URL: http://islamic.utm.my/mansoureh/
URL: http://afes-press-books.de/html/APESS_29.htm

S. K. Gassama
e-mail: skgassama2012@gmail.com

K. Yusoff
e-mail: y.kamaruzaman@utm.my; zuryati@yahoo.com

M. Ebrahimi et al. (eds.), *The Dynamics of Iranian Borders*,
https://doi.org/10.1007/978-3-319-89836-0_1

Over the years, imperialism played itself out in stages. Apparently, European nations had reached the '*limits of their internal agricultural frontiers*' in view of extant practices and prices; thus precipitating the quest for new lands during the fifteenth century to '*plant productive enterprises*' and maximise tax revenue (Kearns 1993: 10). Spain led early imperialism after expelling its Muslims and recolonizing al-Andalus. They established plantations in the Canary Islands and led initial Mediterranean and American despoliation. 'This new imperialist machine was soon exported to Latin America where it served to suck out the silver that oiled European trade with South East Asia [India and China] and brought Asian spices and tea back to stimulate jaded European palates.' According to Kearns (1993), the next stage of imperialism marked the establishment of trading posts by seventeenth-century England and Holland with very 'little control over these societies' (Taylor 1993). These imperial ventures opened an era of mercantile capitalism that enabled European powers to establish monopolies over terms for the exchange for good and services while engaged in forced land grabbing from indigenes. In the Far East and India, the British eventually intervened in political structures and tax-collection. These impositions led to occupation and direct colonization by the nineteenth century. The process witnessed deadly rivalries and global wars as colonial powers projected respective powers and influences over stolen resources. This provides a snapshot of imperial proclivities as a backdrop for the British versus Russian rivalry over Persian territories and her neighbours.

Iran's unique geographical location became a principal cause of contention between great powers at different times. Political relations were often complex and characterised by machinations, intrigues, conflicts and wars, as well as peacemaking treaties and sundry bluster-filled covenants. The authors examine these turbulent periods from the sixteenth through to the third quarter of the twenty century (1970s), relating threats to autonomy via treaties as well as delimitations and demarcations of borders by the great powers, especially Russia and Britain.

Britain achieved its objectives by combining diplomacy with 'divide and rule' cunning; thus forestalling amicable relations between Iran and its neighbours, including Russia. Britain dispatched Sir John Malcolm to woo Iran with the express purpose of preventing an impending French occupation of parts of India under Napoleon. The British Mission was successful and led to the 1801 conclusion of the first ever political and commercial treaty with Iran (see page 18–19). The authors indicate that the principal motive for this and subsequent nineteenth-century treaties was to prevent any Russian advance and to preclude all friendly cooperation between Iran and its neighbours that might jeopardise the strategic objective of securing Britain's territorial ambitions in India.

Following that, authors present similar themes by examining conflicts over Iranian borders beginning with the establishment of the sixteenth-century Shi'ite Safavid State against the express wishes of the Sunni Ottoman Caliphate. Their key controversy was over theological doctrine, but including access to and control of traditional religious sites rather than colonisation. Both states fought several battles and signed treaties, beginning with the Amasieh Treaty of 1555 that ended a twenty-year war. Its provisions included Shi'ite rights to safe-pilgrimage to holy places, such as Mecca and Medina. In addition, Azerbaijan, Eastern Armenia and Georgia were ceded to Iran while Western Armenia and Georgia (up to Shatt al-Arab) were placed under Ottoman suzerainty. The Treaty of Zuhab (1639) made the very first attempt to demarcate a border between the Ottoman Empire and Iran. However, it did not specify which areas belonged to which party. Nevertheless, this treaty ceded Iraq to the Ottomans until the end of WWI, but even then borders were non-definitive. After the Ottoman-Persian war of 1743, the Treaty of Kerden (1746) maintained provisions of the Zuhab Treaty and the Ottomans officially recognised Iranian Shi'ites as Muslims without discrimination.

Although the Kerden treaty ushered in an era of peace, matters were disturbed by the expansionist ambitions of both Russia and Britain, whose objectives were motivated by geopolitical considerations. Iran's strategic location 'permits ready transit of goods from Asia to Europe via large ports with strategic access to the Strait of Hormuz and the Sea of Oman' (see page 27). The significance of Iranian territories steadily increased into the early twentieth century, especially with the discovery of oil and gas. Ongoing machinations of Russia (a territorial power), and Britain (a naval power), only worsened relations between the Ottomans and Persia. In 1622 the Safavid Shah Abbas successfully drove the Portuguese from Hormuz Island with the support of British warships. Even though Shah Abbas afterwards balanced power by allowing the Portuguese to operate castles and trading centres in Kong Port (Bandar-e Kong), the British capitalised on the situation and gradually overshadowed them, becoming entrenched in Iran and the Persian Gulf.

With the death of Shah Abbas, Iran's position weakened and the British took the opportunity to exert more influence for strategically significant 'trade with its Asian colonies and a fulcrum for manipulating the Ottomans' (see page 30). Hence, they established an embassy in Baghdad and a consulate in Basra, and then deployed naval forces to protect the waterways of Shatt al-Arab. They also obtained exclusive rights to navigate the Tigris and Euphrates, supported by an Edict of the Ottoman Sultan.

The Treaty of Turkmenchay (1828) granted Russian influence in Iran as a concession. Although Britain cooperated with Russia against the French on the back of Napoleon's threat to attack British India, they remained apprehensive of Russian expansionism and influence in the Persian Gulf and nearby. Nevertheless, rather than open war, Britain chose to engage in political cooperation with Russia. Hence, their political and military cooperation over Iran and the Ottomans continued throughout the nineteenth century up to 1914. Both imperial powers eyed the exploitation of mineral resources (the British owned oil fields and the Abadan refinery) (Ebrahimi 2016). The British, apprehensive of growing Russian influence in Iran, decided to keep a base in Iraq. A secret agreement was then drawn up to divide Iran into two spheres of influence: the north under Russia and the south under Britain. However, knowing this Russian-British Balkanisation plan, Iran attempted to facilitate German interest as an escape. Ongoing issues forced the precipitation of a resolution at the Permanent Court in The Hague, which then led to the Istanbul Treaty of 1913 and further border redefinitions in 1914 (see page 34).

Regarding borders between the Ottoman Empire and Iran, Russia wished to protect its influence but the British were willing to barter with the Ottomans on account of their oil franchises. Under British-Russian influence and pressure, the Tehran Treaty of 1911 allocated the entire Shatt al-Arab to the Ottomans, although both Iran and the Ottomans had jointly administered the common border previously. WWI witnessed a further realignment of borders in the Persian Gulf with the Shatt al-Arab put under the complete control of the British, along with other sweeping changes.

The Sykes-Picot Agreement of 1916 awarded Britain direct control over Basra and Baghdad, as well as a 'zone of influence' over most of the rest of southern and central Iraq. The status of the northern province of Mosul remained contested until after the war (Agoston and Masters, *Encyclopaedia of the Ottoman Empire* 2009: 283).

Following the creation of Iraq as a British protectorate under a Mandate by The League of Nations, Britain forced Iran to recognise a new nation for which she would also mediate '*reasonable demands*'. Nevertheless, such promises were never delivered, whereupon Iran rejected the 1914 treaty, which then began a new wave of border conflicts between Iran and Iraq with looming war. Iran insisted on Iraq's recognition of the Shatt al-Arab thalweg as the borderline. King Abdulaziz al-Saud of Saudi Arabia began mediation but suddenly passed away. The matter was referred to the International Council in May 1935 but was inconclusive pending further discussions between the both parties. The British initially wanted Iraq to have full control over the Shatt al-Arab, which would incidentally fall under British control and influence. However, knowing this was highly unlikely, they sought participation in joint maintenance of the waterway. Iran refused any joint management that included Britain. Iraq would not accept any joint maintenance commission without British participation.

Iran signed the Border treaty of 1937 and accepted the Protocol of 4 November 1913, as well as the '1914 min' related to border definitions. Key provisions of the 1937 Treaty included acknowledgement of the Thalweg Rule for the Shatt al-Arab (limited to a small area facing Abadan), stipulating that the waterway would be open to all commercial ships with Iraqi renunciation of exclusive right of possession. The 1937 Treaty was not fully implemented and the relationship between both countries deteriorated with the ascendency of Abd al-Karim Qasim to power. Iraqi refusal to allow Iranian officers to guide ships along the waterway then led to tensions and the eventual closure of the Abadan refinery in 1961. Following discussions and mediation, a 1975 Treaty was signed that defined common borders and quieted a long simmering conflict: '... concerning inalterability of the borders and respect for territorial integrity, both sides confirm that their land and river boundaries will be inalterable, permanent, and definite' (Valdani 1998)—until unilateral abrogation by Saddam Hussein in the run-up to the Iran-Iraq war (1980–1988).

Meanwhile, and shifting gears, British fear of Russian expansionism has been a mythically re-echoed theme of geopolitical literature. A twentieth-century British political geographer, Mackinder, known for his concern for the future of the British Empire, gathered the attention of Kevin R. Cox (et al.) who observed:

> The influence of 'geographic conditions' on the power of states was a common analytical concern. This involved an emphasis on environmental variations and territorial configurations, such as Mackinder's concern with the distribution of landmasses, and the ocean in relation to global political power. (Cox et al. 2008: 3)

Mackinder theorised: 'Who rules East Europe commands the Heartland: who rules the Heartland commands the world-islands: Who rules the world-island commands the world' (1919: 19) (Cox et al. 2008: 7). Thus, he addressed the issue of imperial completion by 'setting out his theory of the Heartland and how this drew upon economic history to establish guidelines for defending British imperial advantage' (Kearns 2009: 12). The theory presupposes that the strategic geographical location and size of the Heartland (i.e. Russian Empire) is equal to the might of British sea power. Mackinder's theory thus noted a linkage between geography and imperial power. Mackinder, often described as the '*imperial geographer*' (Kearns 2009),[2] noted 'the mobilization of land-based resources [the Heartland] via the railway looked set to guarantee a redress [of] the balance between sea [the World-Island] and land-power in favour of the latter' (Kearns 2009). Hence, Britain, being small, could only compete for power and glory by keeping its prized colonies from the prying eyes of the Russian Empire.[3]

Hence, Britain's 1801 treaty with Fath-Ali Shah Qajar was not only born of geopolitical considerations to check Russia but, likewise, was delivered to forestall mutual cooperation between France and Iran. The treaty also created an opportunity for Iran to invade Afghanistan, in which case the British undertook to defend Iran in case of attack by Afghanistan or France. However, the Franco-Persian alliance, concluded by the Finckenstein Treaty of 4 May 1807, had aroused so much British concern that they quickly dispatched a diplomatic mission comprising Sir John Malcolm and Sir Harford Jones. The British objective to '*Save India at all costs*' led to the signing of the 1809 Treaty that cancelled all previous Treaties with western governments and promised a stable relationship between Britain and Iran. Iran had been tempted to cooperate with France on Napoleon's promise to do '*all in his power to force Russia's evacuation of Georgia and Persian territory*.' A key provision of the new treaty was the Shah's pledge not to endanger any British interest in India and that Iran would come to the aid of India in case of external attack from Afghanistan or other countries. Britain agreed not to interfere in the conflict with Afghanistan without an invitation from both parties.

Keeping the Russians in check, Britain mediated between Russia and Iran during the war and concluded the Treaty of Gulistan (1813), in which Iran suffered a huge loss of territory (Azerbaijan, Dagestan and eastern Georgia) to Russia. The Turk-

---

[2]Mackinder was associated with the founding of the Geographical Association in 1893, with the objective of familiarizing 'Britain's young with distant colonial prospects.'

[3]Mackinder's Heartland Theory, as propounded in the twentieth century, predicted the waning of Britain as the world island sea power unless Britain checked the rising power of the Russia. It should be noted that by the late nineteenth century, with Germany's desire to create a super naval power, Britain shifted its focus to Germany, which was perceived as an unnecessary luxury; therefore, it was not surprising that naval power proved a battleground for supremacy among European powers during WWI.

menchay treaty of 1828 required Iran to stop any European advance towards India and its ports, and the British guaranteed neutrality in Iranian conflicts with other parties unless solicited by both parties to intervene.

Pursuing the 'divide and rule' paradigm, Britain changed its friendly attitude towards Iran by directly engaging with Emirs in Afghanistan and Central Asia in 1815. They actually instigated the war between Iran and Herat in Afghanistan, leading to the defeat of the latter. Similarly, Britain switched alliances on the notion that 'the end justifies the means' as a major component of its neo-realist analysis of international relations, which became quite evident in central Asia. Fearing her power in 'The Heartland' would likely be reduced by war in central Asian states, Britain reneged and did not aid Iran in its war with Russia. Russia defeated Iran. That led to the Treaty of Turkmenchay (1828), which actually helped achieve another British objective by securing British India from any threat from Iran and Afghanistan.

With Iran's defeat by Russia in 1828, the British immediately became imperial bully and exercised power against Iran by threatening war if the latter's ten-month siege of Herat was not lifted. In his discussion of 'geographical modes of thought... in political decisions', Peter Slowe identified five sources of political power, namely: power through 'might' (aggression and conquest); power through 'right' (mobilisation of people who feel they have a right to territory); 'nationhood' (unique power of the nation state); 'legality' (legal distribution of power and territory); and 'legitimacy' (reality of hegemonic power within and between states) (Slowe 1990: 1–2). Hence, Britain applied power through 'might' and Iran capitulated after the British capture of Khark Island in 1838.

The prevailing power relationships during this second stage of imperialism (Gerry Kearns' frame of reference 1993) exhibited classical realism where 'might is right' and also unencumbered by moral or legal restraints. Classical realists presuppose that the pursuit of power 'at all costs' is a justifiable political goal. Thus, they pursue it by trampling on the interests and rights of weaker states without the least regard to fair play (McGowan et al. 1999: 24). In 430 BC, Thucydides, the father of classical realism, offered this same view of Athenians in *The Medlian Dialogue*,[4] in which Athenians responded to Melians as follows:

[4]In the legendary dialogue between Athens and Melos in 430 BC, the former declared that the latter's purported 'neutrality' during a conflict between Athens and Sparta was tantamount to supporting Sparta and said that Melos would therefore be conquered by Athens. Melians were given the choice of abandoning their allegiance to Sparta and paying tribute to Athens, or facing the consequences if they failed to comply. Thucydides described this dialogue in his timeless masterpiece, *The History of Peloponnesian War.* Considered the epitome of realist thought, it demonstrates that states engage in politics for the sole objective of self-aggrandisement and power over others. The Athenians made what they considered to be a generous offer to the Melians but threatened to use raw power against them if it wasn't accepted: 'the standard of justice depends on the equality of power to compel ... in fact, the strong do what they have the power to do and the weak accept what they have to accept.' The Melians rejected the Athenian position and appealed to their sense of justice, expecting the rule of law would allow the exercise of mercy on a weak state: '[Athenians] should not destroy a principle that is to the general good of all men, namely, that in the case of all who fall into danger there should be such a thing as fair play and just dealing...' (Unoki 2016: 10). The Melians argued that by believing in what they thought was right, God would save them, to which Athens retorted: 'It is a general and necessary law of nature to rule wherever one can ... and we know that you

> Instead we recommend that you should try to get what it is possible for you to get, taking into consideration what we both really do think; since you know as well as we do that when these matters are discussed by practical people, the standard of justice depends on the equality of power to compel and that in fact the strong do what they have the power to do and the weak accept what they have to accept. (McGowan et al. 1999: 24; Unoki 2016: 9)

The Classical concept of power relations, thus portrayed, corresponds well with the prevailing mid-nineteenth century Central Asian milieu. Hence, the British were not the least bit bothered about using might to get what they wanted. Britain was an unrivalled naval power and had the military capability to employ might to secure its geopolitical objectives. In pursuance of these objectives, the British declared war on Afghanistan in 1842 during which they lost 16,500 people (12,000 civilians and 4500 soldiers) in one famous engagement. The Afghans spared only Dr. Brydon to report the catastrophe to his authorities. This devastating war was consequent to intense Russian-British rivalry for hegemony in Central Asia.

The deadly Russian-British rivalry sparked yet another war in the Crimea (October 1853–February 1856). Alliances were also a pronounced feature of European interstate relations from the early nineteenth century, manifesting various approaches that were subsequently exported to colonial efforts during and after the Vienna settlement (Kissinger 1994: 100). Iran entered into an Alliance with the Ottomans in addition to a secret pact with Russia out of fear of British might and influence. The British retaliated by arming and supporting Afghanistan against Iran in the conflict over Herat, which Britain deemed a strategic gateway to India. Thus, out of self-interest, Britain supported the Afghans until 1842, after which they became 'friends in need' in the divide and rule game-play of classical realists. Iran lost Herat to Afghanistan in 1837 when Britain supported the Afghan uprising against Iranian dominion in Herat. This was in clear contravention of Britain's 1814 Treaty with Iran, which stipulated neutrality except on invitation of both parties for British mediation.

Russia and Britain used deceptive tactics to achieve imperial objectives in treaty negotiations, such were the underhanded dealings between Russian and British during the second Treaty of Erzerum (1847), which Iran was duped into signing against their wishes. Similarly, in 1907, Britain and Russia, in spite of bitter rivalry, cooperated to sign the Anglo-Russian Convention that secretly divided Iran into two spheres of influence like war booty (Wright 1977: 30).

Britain saw itself as the guarantor of peace in the region after the signature of the Paris Treaty and thereby sought to reduce Iranian influence by forcing them to evacuate Herat and respect the territorial integrity and independence of Afghanistan. Iranians felt betrayed by the British after Herat's breakaway, a fact expressed in a letter to Lord Palmerston, the British Prime Minister, by Henry Willock, Direc-

---

or anybody else with the same power as ours would be acting in precisely the same way' (Unoki 2016: 9). The refusal of the Melians to denounce utopian idealism and accept Athenian realism caused the latter to condemn the Melian position as a 'lack of common sense' and a 'false sense of honour'. The aftermath was the siege of Melos by Athens, mass slaughter of their men and enslavement of their women in a cold-blooded manner. Thucydides observed that powerful states like Athens compel weaker states to follow their perspectives and dictates or face the consequences of not doing so.

tor of the East India Company in London: 'We have professed to be her [Persia's] protector… we have now become aggressors, upon a more slight pretext than ever Russian encroachment was founded on' (Wright 1977: 59). The Paris Treaty enabled the Balkanisation of the Persian Empire, which then led to the 'political separation of Iran and Afghanistan from their ancient Persian fealty' (see page 23–24) and the historic creation of the state of Afghanistan. While the Paris Treaty declared the independence of Afghanistan and Herat, free from Persian interference (official withdrawal of the Shah of Persia's sovereignty), the British continued to redefine Iranian borders in Khorasan, Sistan and Baluchistan. Major General Goldsmith's unilateral delineation of Iranian borders through the 'First Arbitration' of 1863 and 'Second Arbitration' of 1872 soon followed. The effects of these imperial machinations (border delineations) have left several unresolved disputes to this day.

Furthermore, the authors examine the evolution of post-Soviet Caspian Sea legal regimes related to potential border delineations with respect to exploration and exploitation of a so-called 'common pool' of shared energy and sundry biological resources in the sea. Mackinder's Heartland versus World Island perspective is less applicable in this era because the machinery of the British colonial empire formally ended when devastating World Wars between European colonizers allowed the US to emerge as the Western super power. America most certainly replaced Britain as the neo-colonial state in pursuit of global dominance (including the Caspian Sea), both during the Cold War era and afterwards.

The Caspian Sea was designated the 'Iranian-Soviet Sea' by treaties in 1921 and 1940, which stipulated joint maritime territories. The situation changed after the collapse of the Soviet Union and the emergence of Azerbaijan, Kazakhstan, and Turkmenistan as the Commonwealth of Independent States (CIS). This increased Caspian littoral states from two to five countries, which then necessitated a new legal regime to regulate the Caspian Sea.[5]

Historically, eighteenth-century (1723) Russia had annexed the northern Iranian border along the Caspian Sea after an Afghan insurgency, but withdrew from occupied territory when Nader Shah Afshar became the Persian Regent in 1732. Following a tense relationship with Russia, Agha Mohammad Khan then ceded several territories to Russia, including parts of Dagestan, Georgia, Baku, Darband, Shirvan, Karabakh, Sheki, Ganja, Muqan and upper Talesh, which included Caspian coastlines. Following Iran's defeat in the first Ten Year War, Iran signed the Gulistan Treaty of 1813, which marked the beginning of a Caspian Sea legal regime. After a second Ten Year War, the Turkmenchay Treaty of 1828 ceded even more territories to Russia, including Iravan, Nakhjavan and parts of the Muqan plain, as well as the southern coast of the Caspian Sea. Iran was forced to pay five million Toman as war reparations and the treaty delineated rights to the Caspian Sea to both nations, with severe limitations imposed on Iran.

Iran and Russia agreed that the Caspian Sea should be closed to the rest of the world and shared exclusively by them as a jointly administered inland water asset.

[5]The Caspian Sea was designated a 'Sea' by the 1982 UN Convention on Laws of the Seas and meets the requirements of a 'Closed Sea' under International Law.

In spite of this concession, Iran was denied any military presence in the Sea and Russia never fully respected Iranian rights to it. Hence, the so-called legal regime was described as 'equality in opinion and inequality in practice.' With the collapse of the Soviet Union and the arrival of newly independent CIS countries, a new legal regime was needed. Iran was thus provided with a golden opportunity to reclaim full rights. Nevertheless, issues became more complicated than before, especially after Azerbaijan offered oil concessions to Western companies, although each new state also entered into bilateral agreements with Russia with regard to exploration of resources.

Despite several agreements, all five littoral governments remained in conflict over Caspian maritime borders from 2002 onwards and failed to design a common framework for sustainable exploration of the basin. Consequently, militarisation of the Caspian Sea became a reality despite the existence of the Caspian Sea Cooperation Organisation, which did little to relieve tensions. Various approaches were on offer to define borders. These ranged from different percentage allocations to division according to coastline, as well as common exploitation with a common regime and division into five equal parts. Attempts at bilateral divisions also failed due to key disagreements over off-field resources.

Azerbaijan is described as the 'Caspian Sea legal regime disrupter' because it signed several contracts with Western oil companies to maximise national interests, and hence called for an equitable distribution of borders. Azerbaijan completely ignored existing legal regimes and entered into oil contracts with Western companies in September 1994. They claimed the Caspian Sea was an International Sea and declared that the treaties of 1921 and 1940 were no longer valid or acceptable. It therefore appeared that Azerbaijan had acquiesced to the American bid to make the new Caspian Sea a legal regime, thus enabling a 'divide and rule' strategy by the West.

The US Ministry of Energy and International Monetary Fund (IMF) projected hydrocarbon energy reserves in the Caspian Sea Basin at 178 billion barrels of oil, with an estimated 95 billion located off Kazakhstan, and 33 and 31 billion barrels in Turkmenistan and Azerbaijan, respectively. Gas reserves were estimated at 8.9 trillion cubic feet in Turkmenistan (see page 63). Naturally, this attracted American and global attention, as everyone seeks affordable and reliable energy sources removed from the conflict-ridden Middle East, especially after the 2001 attack on the World Trade Center.

The US clearly seeks the expansion of its influence on the frontiers of its former Cold War rival and desires access to the region's alternative energy resources. Congress adopted a 'Silk Road' Strategy in 1999 to increase its role in the Caspian Sea Basin by promoting free market trading for American companies and friendly relations with CIS Republics. This effort was accompanied by 'rule of law' pretensions and opposition to militancy with the development of stable democratic, Westernised nations. The US also labelled its access to hydrocarbon energy resources in the Caspian Basin and Caucasus regions as 'vital interest' in 1996 (see page 64–65), which automatically brought the entire impetus under military auspices, meaning military options could be deployed to achieve strategic objectives 'by any means necessary'. US foreign policy, since the end of WWII, has clearly been influenced by neo-realist thinking, which further suggests mounting potential for war in the region.

America pursues a 'divide and rule' policy by promoting democratisation in CIS countries with a key focus on reducing Russian, Iranian and Chinese influence and power in the region by bringing former Soviet states into NATO or the East-West Corridor through programmes like 'Partnership for Peace'. Washington even attempted to resolve disputes between Azerbaijan, Turkmenistan and Kazakhstan in 1998. But in spite of key US interests, in traditional geopolitical parlance, the Caspian Basin is Russia's 'backyard' and remains an exclusive sphere of influence with extremely strong political, geographic, economic, military, historical and cultural roots. Russia does not take kindly to any foreign interference in its 'Closed Sea' because such impositions represent major security risks to its survival.

Mid-nineteenth-century Russian-British rivalry in Central Asia most certainly fit Mackinder's framework of imperial hegemony. The Russian Heartland struggled to fill the vast expanse of Eurasia while the British 'World Island' dispersed its sea power via colonies across the globe. Kissinger (1994) noted: 'Great Britain spent the better part of the century attempting to check Russian expansion into Persia and on the approaches to Constantinople and India' (Wright 1977: 100).

By the late nineteenth century, the international competition and struggle between The Heartland and the Island of the World for colonial expansionism extended to the Caspian Sea region (the Great Eastern Sea),[6] an area endowed with huge deposits of oil and gas resources. The authors examine various issues surrounding these reserves and the Caspian Sea in Chap. 5. Oil extraction began in Azerbaijan during the late nineteenth century and accounted for fifty per cent of global production by 1901, which clearly demonstrates the strategic importance of the Caspian region. British and Russian companies competed in the Caspian for this trade. British-based companies included the Caspian Partnership, Bolting Trading and Co., British Petroleum and others. Russian companies included the Russian General Oil Corporation (Zonn et al. 2010: 204).

This imperial rivalry in the Caucasus and Central Asia has been dubbed 'The Great Game', an expression that connotes the struggle to win power, glory and colonial resources at any cost by taking the 'ends justify the means' approach, which is classical realism. The term was first used by the British novelist Rudyard Kipling in his book, *Kim* (1901), to describe the rivalry between Britain, Russia and Turkey for spheres of influence in these regions.[7] Several incidents occurred between Russia

---

[6]Early records showing the inscription of Assyrian King Adadnerrari III date to the eighth and ninth centuries BCE. They read: 'The Great Sea of Sunrise' (or *The Great Eastern Sea*). See (Zonn et al. 2010: 198).

[7]It was reported that a British secret intelligence officer, Captain Arthur Connolly, who was beheaded in 1846 in Bukhara alongside his compatriot, Captain Charles Stoddart, introduced the term, 'Great Game'. Before his execution he indicated that the Great Game was conducted in Asian territories between Britain and Russia. This scenario was often referred to as 'Great Game I'. It was also used to describe the struggle between both European colonial powers over oil and gas resources in the Caspian Sea (Zonn et al. 2010: 198).

and Britain that perpetuated 'Heartland versus World Island' group-think. Against this backdrop, Russia had annulled the Anglo-Russian treaty of 1734 and prohibited British merchants from traversing Russian territories to deliver commodities in Persia (Zonn et al. 2010: 469), while Britain and Ottoman supported Imam Shamil during his Caucasus War of 1817–1864 (Jihad) against the Russians, thereby hoping to reduce Russian power and influence in the region (Zonn et al. 2010: 152).

Russia instituted its containment policy in 1840 in an attempt to permanently patrol Iranian and Turkmen sea coasts in an apparent response to British activities in Iran and the Middle East (Zonn et al. 2010: 474). While the British redrew Iran's southern borders, they failed to have much influence on Iran's northern borders due to Russian resistance to the World Island's naval power advances into the Heartland. The Crimean War (1853–1856) typified mid-nineteenth-century contests for glory, power and trans-regional domination. This war began between Russia and Turkey on account of the former's crusade to attain 'Protector' status for Orthodox Christians in the Ottoman Caliphate. Addicted to opportunism yet bitter enemies engaged in vicious rivalries and wars elsewhere, Britain and France found room to become partners in classic neo-realist style and deliver defeat to a common enemy they wished confined to the Heartland, minus naval power.

Oil explorations in the Caspian region brought new bouts of rivalry as Russia recovered from its abysmal exit from WWI and re-emerged as a superpower post-WWII. Russia saw Iran as a rival for energy exports to Europe but also as an ally against the Commonwealth of Independent States and Western European powers. Using its power and influence, Russia wished to ensure that all oil and gas exports from the Caspian region (Kazakhstan and Turkmenistan, etc.) would pass through Russia. Nevertheless, in spite of many advantages provided by Iran to Russia, Iran remained marginalised among Caspian powers by American opposition to Iranian participation in regional oil and gas projects due to unresolved post-revolution issues.

Meanwhile, relations between Iran and the European Union were less tenuous in view of EU aims to secure alternative sources of energy. US strategy was to control Caspian Sea energy at the source and aggressively curtail Russian, Chinese and Iranian dominance by proximity. China's increasing demand for hydrocarbon energy irked US policy-makers: 'China will gain increasing significant influence in Central Asia… it is difficult for us to compete with China in some of these countries' (Boroujerdi et al. 2011). Another key regional power is Turkey, whose foreign policy is based upon a pragmatic approach that 'minimises problems with neighbouring countries with a foreign policy in accordance with regional circumstances'. Turkey sees itself as bridge between two gigantic geo-cultural groups: the Islamic Middle East and the secular-progressive West (Terterov et al. 2010). Hence, Turkey cooperated and supported US interests in the Caspian region and also gave leeway to Russia. Turkey is a major gas transit route from the Caspian and hosts various pipelines to Europe.

Iran opposes bilateral agreements between Caspian littoral countries and Western powers because they would undoubtedly lead to wild exploitation of Caspian resources without regard for future generations (Bahgat 2005). Until 2001, there were disputes between Iran and Azerbaijan over the oil fields of Araz, Arav and

Shargh. Hence, Iran used the military and deployed warships supported by air forces to stop Azerbaijan's reconnaissance aircraft working with the British Oil Consortium; thus pre-empting discoveries in the Alov (Alborz field) (Laruelle and Peyrouse 2009: 21).

Similarly, the US and its allies promote the creation of regional organisations and bilateral cooperatives that create ambiguity regarding Caspian borders and pose potential risks to Russian and Iranian interests. Obviously, this generates a continuum of discord for all parties with regard to access and agreements on the southern side of the Caspian Sea, thereby forcing a need for border clarifications between Iran (Russian ally) and Azerbaijan and Turkmenistan (US allies). These matters indicate that no single country has full control of the situation, which makes for a worrisome near future.

The authors then discuss the separation of Bahrain from Iran, one of the most painful of all of Iran's territorial losses to Ottoman, Russia and Britain. Until its separation in 1970, Bahrain had historically been designated the 'fourteenth province of Iran', hence its independence was a bitter pill to swallow. Bahrain's archipelago is strategically located and supported trade and the safe passage of British colonials. Therefore, as a naval power bent on countering the Heartland's expansion into Iran, Central Asia and the Middle East, Britain viewed the Island as a tactical tool that enhanced its military power in support of colonial objectives in the Middle East and Central Asia. With the discovery of oil and gas in 1908, geopolitical factors were only magnified by new economic considerations. Oil had allowed Britain to change its commercial and military fleets from coal-fired steam power to the greater efficiency and power of the diesel engine.

Halford Mackinder's theory of the 'Heartland versus the World Island' was no more evident than in the Middle East, especially in the rivalry for control of Persian Gulf. Thus, a strategic military base in Bahrain served not only British interests in Iran, Central Asia and British India, but also allowed them to prevent the Ottoman Caliphate's premature disintegration, which otherwise would have led to serious economic losses for Britain that would only benefit the Russian Heartland. Moreover, the Tsar wished to be the 'Protector' of all Christians in the Ottoman Empire, something Britain fought tooth and nail to prevent. This Russian aspiration was a key factor in the Crimean War (1853–1856) between Russia and Turkey, as supported by Britain and France to purposely curtail Russian naval power in the Black Sea.[8] Salisbury wrote in 1892: 'The protection of Constantinople from Russian conquest has been the turning point of the policy of this country for at least forty years, and to a certain extent, forty years before that' (Mahajan 2002: 18). The British support of the Ottomans allowed them a window through the Caspian Sea that kept Russia in check. Hence, friendly relations with the Ottomans and their protected areas in the Middle East were of paramount importance to Britain.

By combining the classical realism of 'might' and 'ends justify the means' with diplomacy via deceptive treaties, Britain successfully established its own brand of

---

[8]Encyclopedia Britannica. Crimean War: Eurasian History (1853–1856). Online at https://global.britannica.com/event/Crimean-War (accessed 1 May 2017).

geopolitical/economic power in Bahrain right under Iranian noses. In 1819, Commodore Bruce, Commander of the British Fleet in the Persian Gulf, signed a Pact with Hossein Ali Farmanfarma, the Fars Governor, a provision of which stated '… until the time when the Iranian State cannot protect the Persian Gulf, it is the responsibility of the British Government' (see page 97). This clause obviously indicates that Britain fully intended to engage machinations and deceptions to create exactly that condition. Indeed, the British encouraged Arab nationalism by recognising Bahrain as an Arab State rather than part of the Iranian state. When Iran's parliament declared Bahrain 'an integral part and the fourteenth province of Iran forever' in 1957, London protested by saying Bahrain was '*an independent Arab state*' (see page 110). Unknown to Tehran, Britain had entered into 'private relations' that included agreements and treaties with local rulers in clear contravention of all previous treaties they had signed with Iran about non-interference in Iran's internal affairs. Article 2 of the 1814 Treaty clearly stated: 'When there is conflict between tribal chiefs and border rulers, Britain has no right to interfere, and if rulers offer a part of Iranian land to Britain instead of help, the British government must refuse and not interfere in Iran's internal affairs' (Ashtiyani 2005).

America's entrance into the Middle East melee after WWII added a new dimension to international competition among great powers. As European powers collapsed, US interests emerged more powerfully than British interests in the Middle East, including in the Persian Gulf. Iran's protests fell on deaf ears in the comity of nations as the US continued its support for British colonial impositions through might, especially de-Iranization by rejecting Iranian sovereignty in Bahrain. Iran had little choice other than the farce of a dignified escape through a referendum brokered by British and American influence.[9] Hence, in 1966, Mohammad Reza Shah declared Iran would not seize Bahrain by force but preferred self-determination by Bahrainians under UN supervision. A prejudicial approach was taken to the referendum, which was conducted under restrictions that were limited to consultants who were clearly associated with business groups linked to British interests. On 11 May 1970, the UN Security Council passed Resolution No. 278, which approved Bahraini independence and membership in the UN. Iran reaped an empty hand of international praise for Mohammad Reza Shah's 'toleration and acceptance of democratic principles in Bahrain.' Meanwhile, Bahrain's Parliament passed an Act of Independence on 24 May 1970, with 199 in favour and 4 against. Iran then became the first country to recognise Bahrain as an Independent state, thereby finalising the separation.

[9]In fact, after World War II, the British and Americans used their influence to control most of the political movements of nationalisation as well as efforts in democracy in the Middle Eastern countries by any means. The British plot and American operation to remove Iran's democratically elected Prime Minister from power in 1953 is one of the obvious instances which prove the leverage of both powers in Iran (Ebrahimi and Yusoff 2015).

1.1 Organisation of Chapters

In summary, Iran's strategic location between the Caspian Sea and the Persian Gulf, with access to the Tigris and Euphrates Rivers, made it the primary entry point to Central Asia, the Caucasus and Anatolia known as Asia Minor for European colonizers. Moreover, Iran's rich history and civilisation, plus its long established governance structures, were comparable to those of occidental interlopers. Therefore, the latter manipulated these extant structures (socio-economic and political) by diplomatic means (treaties, deception, 'divide and rule', etc.) as well as by coercion (war and/or occupation) to satiate imperial appetites without direct colonisation. Our authors indicate that during this period of foreign imperial ascendency in Iran, the nation had weak rulers who were easily manipulated and consequently permitted the substantial losses of Persian territory that led to the late twentieth-century reduction now known as Iran.

Chapter 2 investigates 'a series of treaties between Iran and Britain during the nineteenth century that brought about the great loss of Iran's ancient eastern frontiers'. The authors trace the origins and conclusions of numerous treaties with Britain and Russia throughout the nineteenth century, all of which led to demarcations of Iran's eastern boundaries. As such, they wrought a gradual reduction of Iranian territory over time with a strategic British objective not so much to aggrandise Iranian territory as to protect and defend 'British India', the '*Jewel in the British Crown*'. Hence, Britain did not wish heavy involvement in Iran but rather to nip the bud of any Russian attempt to expand and control Iranian territory.

Chapter 3 focuses on the origins of conflict between Iran and Iraq. It concerns root causes of conflict between Iran and Iraq and presents processional facts regarding boundary formations vis-à-vis influences of western powers. The chapter provides a detailed explanation of Iran's historical dominion of what is now Iraq, and moves forward to the establishment of the Ottoman Empire and modern disputes between Iran and the Ottomans. Again, European influence (Britain and Russia) on Iran-Ottoman relations served to enhance these disputes, only to fuel successive wars that resulted in unpleasant treaties for the Iranian cause. A continuum of British interference ensued that favoured the establishment and eventual independence of Iraq under Britain's tutelage.

The Fourth Chapter examines Caspian Sea Legal Regimes and reviews the history and evolution of the legal regime around the Caspian Sea after the Soviet Union's collapse. The Caspian Sea region has always been a focus of political strife involving Russia, Kazakhstan, Turkmenistan, Azerbaijan and Iran. None of these has yet to reach any settlement regarding boundary lines in the Caspian Sea. This area is, therefore, a matter of great concern for the United States of America as well, due to its rich natural resources. Nevertheless, not only regional attitudes of the Sea's natural neighbours, but also of America have prevented any resolution.

The discussion flows into Chapter Five's survey of post-Soviet Legal Regimes. The chapter describes the Caspian Sea's energy resources and traces the roles of regional and trans-regional powers regarding the area's legal regime. The most important energy resources involve incredibly large estimates of oil and natural gas reserves that only invite the biased interests of international and regional powers. The competition has more than a decade's history of negotiations with no solution in sight. Despite its rich resources, the absence of a specified legal regime as submitted by major powers only increases the risk of armed conflict for littoral Caspian countries.

Chapter 6 discusses Britain's cunning separation of Bahrain from Iran. It clarifies the history of Bahrain's separation from Iran and analyses western influences, especially British policies in the region. Suitably situated in the Persian Gulf, Bahrain was yet another traditional part of Iran's vast territories. British expansion in Bahrain and its empowerment of local tribes in the face of a weakened Iranian government provided fertile grounds that stabilised British dominance in the region. Economic interests, oil, and the matter of India gave the British cause to consolidate its monopoly in Bahrain to the exclusion of other European powers. Contracts with Bahrainian Sheikhs ignored Iran's central government and paved the way for British success. Consequently, British hegemony, together with other factors, brought about Bahrain's independence and separation from Iran. We conclude the discussion in the last chapter. All seven chapters are linked by various themes and sombre trials that confronted Iran as the nation attempted to protect and retain pliable borders from extrinsic impositions by imperial Russia, Britain, Portugal and Ottomans.

References

Agoston, G., & Masters, B. (2009). *Encyclopaedia of the Ottoman Empire*. New York: Facts On File, Inc.

Ashtiyani, A. I. (2005). *Tarikh-e Iran az Sadr-e Islam ta Enqeraz-e Qajariyye (Iran's history from early Islam to the collapse of the Qajar Dynasty)*. Tehran: Samt Publications.

Bahgat, G. (2005). Oil and terrorism: Central Asia and the Caucasus. *The Journal of Social, Political, and Economic Studies, 30*(3), 265–281.

Boroujerdi, A., Ansari-Zadeh, S., & Karami, M. (2011). Zheoeconomy-e Darya-ye Khazar va Ta'sir-e an bar Amniat-e Enerzhi-ye Etehadiye-ye Ouropa (The geoeconomy of the caspian sea and its impact on the energy security of the European Union). *Faslname-ye Rahbord (Journal of Rahbord), 20*(60), 7.

Cox, K. R., Low, M., & Robinson, J. (2008). Introduction: Political geography: Traditions and turns. In: K. R. Cox, M. Low, & J. Robinson (Eds.), *The SAGE handbook of political geography.* London: SAGE Publications.

Ebrahimi, M. (2016). *The British role in Iranian domestic politics (1951–1953)*. Cham: Springer.

Ebrahimi, M., & Yusoff, K. (2015). The British plot to remove Iran's democratically elected Prime Minister from power. *Acta Histriae, 23*(4), 735–748.

Encyclopedia Britannica. Crimean War: Eurasian History (1853–1856). At: https://global.britannica.com/event/Crimean-War (May 1, 2017).

Kearns, G. (1993). *Fin de siècle* geopolitics: Mackinder, hobson and theories of global closure. In P. J. Taylor (Ed.), *Political geography of the twentieth century.* London: Belhaven Press.

Kearns, G. (2009). *Geopolitics and empire: The legacy of Halford Mackinder*. New York: Oxford University Press.

Kissinger, H. (1994). *Diplomacy*. New York: Simon and Schuster.

Laruelle, M., & Peyrouse, S. (2009). The militarization of the Caspian Sea: 'Great Games' and 'Small Games' Over the Caspian Fleets. *China and Eurasia Forum Quarterly, 7*(2), 17–35.

Mahajan, S. (2002). *British foreign policy 1874–1914: The role of India*. London: Routledge Taylor & Francis.
McGowan, P. J., Cornelissen, S., & Nel, P. (Eds.). (1999). *Power, wealth and global equity: An international relations textbook for Africa*. Lansdowne: University of Capetown Press.
Slowe, P. M. (1990). *Geography and political power: The geography of nations and states*. London: Routledge.
Terterov, M., Pool, J. V., & Nagornyy, S. (2010). Russian geopolitical power in the Black and Caspian Seas region: Implications for Turkey and the World. *Insight Turkey, 12*(3), 24–25.
Taylor, P. J. (Ed.). (1993). *Political geography of the twentieth century*. London: Belhaven Press.
Unoki, K. (2016). *International Relations and the origins of the pacific war*. Basingstoke: Palgrave Macmillan UK.
Valdani, A. J. (1998). Barresi-ye Ahdname-ye 1975, Iran va Iraq (Review of 1975 Iran-Iraq Treaty). *Mahnameh Itila'at-e Siyasi Ightisadi (Monthly Politics-Economy Information), 247–248*, 40–55.
Wright, D. (1977). *The English amongst the Persians: Imperial lives in nineteenth-century Iran*. London: IB Tauris.
Zonn, I. S., Kosarev, A. N., Glantz, M. H., & Kostianoy, A. G. (2010). *The Caspian Sea encyclopedia*. Heidelberg-Berlin: Springer-Verlag.

Mansoureh Ebrahimi (Iran/Malaysia) Ph.D., is a senior lecturer with the Faculty of Islamic Civilization, Universiti Teknologi Malaysia. Dr. Ebrahimi has written about Iran's culture, history, religious beliefs and Malaysia's *halal* industry from the Islamic perspective. She also teaches Persian and Philosophy of Science and Civilization. Her books include *The British Role in Iranian Domestic Politics (1951–1953)* (Germany: Springer Nature, 2016); Mansoureh Ebrahimi and Kamaruzaman Yusoff (Eds.): *The Halal Industry in Malaysia: Prospects and Challenges* (Malaysia: UTM Press, 2015); has also contributed several articles on culture, religion and history (including Malaysia, Middle East, Iran and Turkey) to reputable peer-reviewed journals; 'The British plot to remove Iran's democratically elected Prime Minister from power', in: *Acta Histriae* (Republic of Slovenia: Annales Publishing House), 23, (2015): 735–748.

Saikou Kawsu Gassama (Malaysia) is a Ph.D. student in the Faculty of Islamic Civilization, Universiti Teknologi Malaysia. He holds a B.A. in Political Science from the International Islamic University Malaysia, an M.A. in Islamic Political Economy from the University of Durham (2001), an M.A. in Social Science from the National University of Malaysia (2004), and an M.A. in Diplomacy, Law and Global Change from Coventry University (2007). He has worked as a Programme Specialist Governance and Partnership at Action Aid International in The Gambia (Sept 2013–Sept 2016), as a Lecturer at the Department of Political Science, School of Art and Sciences, The University of The Gambia, and served as the Secretary to the Cabinet, Office of the President of the Republic of the Gambia (Mar 2011–Mar 2012), and Permanent Secretary 1, at the Office of the President (Feb 2010–Mar 2011).

Kamaruzaman Yusoff (Malaysia) is professor at the Faculty of Islamic Civilization, Universiti Teknologi Malaysia, Head of West Asia Civilization Research Group, where he specialises in Middle Eastern History and Islamic Studies. He holds a B.A. in History from the National University of Malaysia; and an M.Litt. and a D.Litt. from the Department of Islamic and Middle Eastern Studies, Edinburgh University, Scotland. He is the author of several articles and books on the Middle East, Orientalism in Islamic History, and the relationship between the Ottoman Empire and the Malay World; and has served as Head of Department Research Group and Dean of the Faculty.

Chapter 2
On Iran's Eastern Borders: Origins and Consecutive Treaties with the British During the Nineteenth Century

Mansoureh Ebrahimi and Kamaruzaman Yusoff

Abstract This chapter examines various treaties between Iran and Britain during the nineteenth century. It discusses the imposition of unjust treaties by regional and western powers as pacts that eventually separated Herat from Iran, a traditional component of the Persian Empire from ancient days. Moreover, Iranians, though not intimate with Afghans, nevertheless regarded their people as close tribal relatives with similar languages, religion and culture. Long before Afghanistan independence, and even after it, Afghans paid taxes, ransoms and tributes to the Iranian government. After its long history as part of Iran, Herat became independent by British cunning, and for the first time in its modern history, Afghanistan became an independent nation by means of British impositions.

Keywords The British · Iran-British treaties · Iran's Eastern borders
Demarcation of Iran's boundaries

2.1 Introduction

Although Britain had no intention of taking any of Iran's territory, it played a crucial role in the demarcation of Iran's boundaries, particularly its eastern borders, which ran north to south from the north-east of Khorasan to Makran beach while passing Afghanistan, Sistan and Baluchistan. These areas had been under Iranian sovereignty for centuries, but in the mid-nineteenth century they became semi-independent states

M. Ebrahimi (✉) · K. Yusoff
Faculty of Islamic Civilization, Universiti Teknologi Malaysia, 81310 Johor Bahru, Johor, Malaysia
e-mail: mansoureh@utm.my; suriebrahimi@gmail.com
URL: https://www.linkedin.com/in/suri-ebrahimi-08aa0457/
URL: http://islamic.utm.my/mansoureh/
URL: http://afes-press-books.de/html/APESS_29.htm

K. Yusoff
e-mail: y.kamaruzaman@utm.my; zuryati@yahoo.com

M. Ebrahimi et al. (eds.), *The Dynamics of Iranian Borders*,
https://doi.org/10.1007/978-3-319-89836-0_2

under local tribal chiefs and governors. Iran's eastern regions are mostly arid, infertile stretches of sand and gravel to which, nevertheless, the proximity of British-India brought much turmoil.

These inhospitable expanses are so vast they posed several sets of difficulties for the British, whose hold on India played a major role in relations with Iran. At the time, maintaining and defending its jewel (India) while impeding Russia's reach were major British objectives (Wright 1977). As a platform of perennial regional conflict, particularly between Iran and Afghanistan, the region witnessed much bloodshed due to British-India intrigue. The usual results were one-sided treaties in which British concern was never 'fear of Iran' but rather of Russian expansion. Britain always considered and protected its own interests rather than regional beneficence. Hence, its unwritten charter authored an indigenous legacy of revolt and aversion to British intervention and hegemony. At the time, Britain wielded much power with which it accomplished a redefinition of regional boundaries that never pleased Iran. By the mid-nineteenth century, Russian ambitions prompted Iranian ire and force for both the sake of honour and displeasure with British interpositions. Nevertheless, this chapter does not examine the formation of Iran's eastern borders but rather investigates a series of treaties that brought about the great loss of Iran's anciently held eastern frontiers.

2.2 Iran-British Treaties

Britain considered Iran a bridge of defence to protect its interests in India from Russia and the French. Hence, British policy purposed to prevent amiable relations between Iran and its neighbours by substituting the ruse of British goodwill as Iran's best and only ally, thus establishing its chief objective for the entirety of the Qajar dynasty. All nineteenth-century treaties between Iran and Britain or between Iran and its neighbours were the result of British efforts to guarantee this very policy. Britain's strategies involved precipitous actions whenever probabilities even suggested a threat to its interests in India, including the dispatch of astute and well-informed representatives to the Shah of Qajar.

2.2.1 Treaty of 1801

The first such probability arose with Napoleon's portentous hint of a French occupation of India via the Iranian corridor, for which Britain dispatched Sir John Malcolm from Mumbai to Bushehr in February 1800. His mission was to negotiate a mutual defence treaty with the Iranian government that also encouraged Iran to invade Afghanistan and thus pre-empt any Iranian cooperation with France while improving Iran's commercial relations with Britain. Sir John Malcolm settled this treaty in 1801 with Fath-Ali Shah Qajar, with the Shah's signature stamped by his chancellor, Haji

Ibrahim Khan Sadr Azam, in addition to government members and the rulers of Fars, Khorasan and Persian Gulf territories, who all promised to cooperate with the British under all circumstances (Mahmoud 1982/1361). This enacted the first political and commercial treaty between the two countries. Politically, Fath-Ali Shah agreed to attack the Afghans if they moved against India, and to prevent the French from stationing troops or remaining in Iran. The British agreed to defend Iran in the event of attack by either the Afghans or the French (Ghani 2013). Economically, Britain promised to provide Iran with artillery, ordnance, essential defence and fieldworks, as well as weapons suites and mercenaries in case of attacks by either Afghanistan or France (Mahmoud 1982/1361).

However, the 1801 Treaty did not sufficiently relieve Britain of its concerns. Surprisingly, Fath-Ali Shah's loyalty to the treaty was steadfast to the extent that the French were never able to sway him. His final answers were simply 'no' to any French delegation. For instance, Colonel Roimeu presented him with Napoleon's letter promising that if Iran cancelled the 1801 Treaty with Britain, Napoleon would provide sufficient troops and financial aid to enable the retaking of Caucasia and Georgia (Kaye 1874).[1] The Shah replied that even if Napoleon were to enter Tehran, Fath-Ali Shah would not meet him (Mahmoud 1982/1361), although later Fath-Ali Shah became disappointed with the lack of promised British assistance. '*France sent General Gardane to Iran, after concluding the Finckenstein Treaty on 4 May 1807, undertaking to restore Georgia to Iran if the Shah joined France in war against Russia*' (Rayfield 2013). This new Franco-Persian friendship caused the British immense concern. Consequently, they sent Sir John Malcolm and Sir Harford Jones (Chief Resident at Basra for the East India Company and Ambassador) to Iran.

2.2.2 *Treaty of 1809*

The festering 'Save India at all costs' British condition had only one remedy: an intimate apposition with the Persian Royal Court. Sir John Malcolm arrived in Bushehr on 10 March 1808, but the Royal Court had been offended by Britain and avoided any accommodation, so he left for Kolkata on 12 July 1808. Britain's political strategies had changed over the long term. Hence, Fath-Ali Shah's sympathies ambled towards Napoleon's promises of using 'all in his power to force Russia's evacuation of Georgia and Persian territory' (Rayfield 2013). Sir Harford Jones went to Iran seeking a compromise between Russia and France in Tilsit (Ghani

[1] After Paul I's death (only son of Peter III and Catherine, Emperor of Russia between 1796 and 1801, who had planned to attack and occupy India, Tsar Alexander I took power and completely changed Russian foreign policy. France became the enemy and good relations with Britain were established. However, Russia desired Iran's Georgia, believing it was Russian territory. Hence, Russian troops invaded Georgia on 12 September 1802. Fath-Ali Shah (Fath-'Ali Shah) asked for British aid but was refused. Finally, in January 1807, the Iranian ambassador to Britain, Muhammad Nabee Khan, returned to Iran without fruitful results.

2013),[2] and also to prevent Iran from joining the dangerous coalition. The result was a treaty of all-embracing unity and accord between Iran and Britain.

On 12 March 1809, Sir Harford Jones, Mirza Muhammad Shafi‘ Mu‘tamid al-Duleh (Grand Vizier), and Haji Muhammad Hossein Khan Amin al-Duleh, delegate of the Persian Majesty, signed the treaty (Hessam-Moezzi 1987/1366). The 1809 Treaty (Treaty of Finckenstein) cancelled all other Iranian treaties while ensuring the defence of British-India. The treaty's first objective mentions the stabilization of sincere and enduring relations. The Shah of Persia was obliged to cancel all former treaties with European governments and further, to prevent the passage of any European troops en route to India. The most notable item was the Shah's pledge to refuse any part in any action or move against the British Royal Majesty or the endangerment of future British interests in India in any manner whatsoever (Hessam-Moezzi 1987/1366). Also, should Afghanistan or any other government invade India; Iran was to take military action to suppress them. In return, Britain provided a yearly *loan* of 200,000 Toman to the Iranian government for the preparation and instruction of its soldiers in the event of war with any invader. This brilliant move facilitated the expulsion all European attack forces (Mojtahed-Zadeh 1996). Moreover, Britain agreed not to interfere in Iran-Afghan conflicts unless both parties desired Britain to mediate. The British government approved the treaty in 1810 and Sir Gore Ouseley was sent to Tehran as ambassador (Mojtahed-Zadeh 1996).

2.2.3 The Treaty of 1814

Soon after Sir Gore Ouseley's arrival in Tehran, the conflict between Russia and Iran came to the boil. Britain wished to end the war and avert even more serious consequences. This was accomplished by the treaty of Gulistan (12 Oct. 1813) between Russia and Iran, mediated by Britain. Nevertheless, it represented a solid defeat for Iran because the entire territory North of the Aras River (Azerbaijan, Dagestan and Eastern Georgia) was taken from Iran, as the river became Iran's new northwest border (Partu 2009/1388).[3] The British were obviously unhappy with their previ-

[2]The Tilsit Treaty of July 1807, signed between France and Russia, initially prevented France helping Iran, and thus annulled Napoleon's plan. When Iran signed the Treaty of Tilsit, the French lost interest and removed their political-military missions from Iran (Ghani 2013), especially as Spain presented yet another barrier. Thus, France could not effectively ally with Iran, who needed assistance.

[3]Russians never gave back Iran's land despite what it had promised in the Gulistan (Gulistān) Treaty. In the case of Iravan (Īravān/Yerevan), the dispute on the Russo-Persian frontier started once again in 1826. Although Iran had become totally weak after many years of war, it finally engaged in war with Russia. It was fourteen years after the Russo-Persian war that eventually ended in the Gulistan Treaty. See more clarification about Gulistan Treaty in Chap. 4 of this book. Iran's army was supported by neither Caucasian's soldiers, nor Azerbaijan's people. At the result, after two years' war, unfortunately Iran's army failed. When conflict and quarrelling began on the Russo-Persian frontier, the British government sent Sir Colonel John Kinneir MacDonald to Iran. Then, on 10 February 1828, the Treaty of Turkmenchay was signed between Iran and Russia. This

ous treaty, which had obligated them to support Iran in the event of any European attack (Ghani 2013). After the Russo-Persian Treaty, Britain managed some changes. For this purpose, a new mediator was required as Sir Gore Ouseley had negotiated the 17 March 1812 Treaty and could not be expected to ask for changes. This left Messrs Muir and Henry Alice to sign a new treaty on 25 November 1814 (Mahmoud 1982/1361).

At the time, Mirza Abul Qasim Farahani was still alive. A clever politician, he did not agree to the same commercial treaty with Britain as had been assigned to Russia with the Turkmenchay Treaty. The treaty of 1814 obligated Iran to block the advance of any European nation towards India and its ports. Its ninth article stated that in case of conflict between Afghans and Iranians, the British would not interfere on behalf of either party unless solicited to arbitrate the peace by both parties (Aitchison 1876).

2.3 British and Iran-Afghanistan Relations (1815–1857)

In 1815, Britain changed its policy towards Iran as they had initiated friendly relations directly with the Emirs of Afghanistan and Central Asia. In other words Britain's ease with the government in Afghanistan caused British indifference towards Persia's Royal Court. Although Iran faced disturbances in Astarabad and Khorasan, the Emir of Kharazm attacked Khorasan and his Iranian troops vanquishing all in their path. He then repeated the heroics in 1817 by subduing Prince Firouz Mirza, the governor of Herat. However, that same year, the Afghan Shah, Mahmud Kabuli and Fatih Khan (his minister) allied with Hezareh and Qaraei tribes to capture Khorasan (Mahmoud 1982/1361). Meanwhile, Muhammad Vali Mirza, the governor of Khorasan, was deposed and his brother, Shuja' al-Saltaneh (governor of Tehran), was appointed his successor. Afterwards, the Afghanis, in particular Shah Mahmud, became frightened and adjusted their posture. The emir of Herat sent his minister with generous offerings and a letter of apology. He presented the key of the castle of Ghorian together with pure gold valued at 50,000 Toman to his Iranian counterpart (Mahmoud 1982/1361). The Afghans evacuated Ghorian, which they had previously captured. Thus, Herat surrendered to Iran and Iran's Prince resolved to place the seal of his power in the eastern and south eastern regions of the country more than ever before.

However, at the same time, Britain was heavily invested in Yangon, Sindh, Sikh and Oude and could not control the region or stop any Iranian advance into Central Asia. Moreover, Russia and the Ottomans were also rattling their impressive sabres. Hence, British policy necessitated the creation of regional war—an ancient ruse used by authorities to maintain hegemony by weakening any nation that obstructs their objectives, enemy and not. After Iran's defeat by Russia leading to the Turkmenchay Treaty of 10 February 1828, Britain was assured that Iranian troops would no more

---

was a political-commercial agreement that led to the cancellation of the Gulistan Treaty, and Iran lost many parts of its northern frontier, including Iravan and Nakhjavan (Nakhchivān). See Partu (2009/1388).

take to the field and that Afghanistan was similarly and thoroughly debilitated, as graciously desired by the British Crown.

At the time, Herat's authorities were subordinate to the Iranian Shah and most civil disturbances in the region quickly ceased with the arrival of an Iranian army. In 1837, Kamran, Shah Mahmud's son, was governor of Herat. He provoked the Emirs of Khorasan and other neighbouring Islamic states. He appeared to stand against Iran, dissociated from Tehran and conducted a military campaign to Farah, Ghorian and Sistan (Mojtahed-Zadeh 1998/1377). At the same time, Muhammad Shah, grandson of Fath-Ali Shah,[4] decided to subdue the eastern governors with a military campaign that stretched from Tehran to Herat. On 22 November 1837, Iran's army, led by Haj Mirza Aghasi, began a ten-month siege of Herat, during which Yar Muhammad Khan offered to surrender. However, British policy would not be compromised. Sir John McNeill, British Minister Plenipotentiary in Tehran, cut political relations with Iran and officially informed the Iranian government that if they did not lift the blockade of Herat, Britain would declare war on Iran. Colonel Stewart delivered the ultimatum to Muhammad Shah. On 25 June 1838 Muhammad Shah arrived in Mashhad, and on 9 September 1838 he broke the siege and removed his army to Mashhad. Iran conceded because British warships had entered the Persian Gulf and captured Khark Island on 7 June 1838 (Mojtahed-Zadeh 1998/1377).

In time, British arrogance accused Iran of befriending Russia, its neighbour, as an excuse to prevent Iran's victory in Herat. The British assumed that if Iran seized Herat the entire region might eventually become Russian territory. However, the reality is/was that Russia was thousands of kilometres from India and could never reach India. Even half a century after Russian troops reached Herat and Mazar-e Sharif, they were still, under no circumstances, prepared to interfere in Afghan affairs (Bahmani Qajar 2007/1385). Indeed, British policy was to relentlessly expand Britain's influence in Central Asia. After these measures were taken, Afghanistan effectively became a British protectorate. Britain occupied Kabul and other cities in the surrounding area. By early 1840, all of Afghanistan was under British dominion. Although Herat's governors had looked to Iran for help, the Shah of Persia had become so cowed by the British he no longer cared. While Shah Shuja'al-Mulk aligned with British power in Afghanistan, both Dost Muhammad Khan in Turkestan and Yar Muhammad Khan in Herat encouraged Afghans to resist the British. On 18 September 1840, a great battle between British troops and Dost Muhammad Khan's army led to the latter's arrest in Kabul on 12 November 1840 (Mahmoud 1982/1361).

In the following year, Yar Muhammad Khan of Herat proclaimed independence, which provoked a successful Afghan uprising, after which, on 1 January 1842, British troops left Kabul after signing a treaty. However, on 06 January 1842, a massive slaughter of British troops occurred as they returned to India. The Afghans killed 16,000 men, leaving only Dr. Brydon to report the utter British failure. Such slaughter was among many consequences of the competition between Britain and Russia in

---

[4]Fath-Ali Shah (1212–50/1797–1834) died on 23 October 1834. His son, the crown prince, Abbās Mīrzā (1203–49/1789–1833), died a year before his father in 1833.

Central Asia. This catastrophe and other failed British strategies to garner support from Emirs in Kabul and Herat effectively brought the region closer to Tehran.

Despite the enormous reversal, British concerns again festered and fostered yet another treaty (25 January 1853) with Iran, this time signed by Mirza Agha Khan Nouri, the grand vizier of Nasir al-Din Shah's reign. It had six objectives (Mahmoud 1982/1361). Accordingly, Iran would (i) never send troops to Herat unless attacked by an external power; (ii) never intervene in the internal affairs of Herat; (iii) desist from minting coins, reading sermons, or any other measures that would induce Herat's submission to Iran; (iv) release every Herat emir held in the prisons of Mashhad, Tehran or elsewhere, unless they intended to remain in Iran and serve the Iranian government, in which case they should be afforded all due consideration for their welfare. Each obligation supported Herat's independence on condition of British neutrality, which otherwise would automatically annul the treaty. It was, however, impossible for Britain to remain loyal to such terms, which, in essence, were exactly the same as recorded in the previous accord.

Meanwhile, Britain's desire to contain Russian expansion stirred yet another war in the Crimea, from October 1853 to February 1856. The Ottoman Empire declared war on Imperial Russia as Britain's anxiety over India caused a Turkish alliance with Iran. However, Iran had signed a secret agreement with Russia in 1854, which the British knew about, subsequently taking action against Russia, for which Iran paid the penalty. The British, believing Herat was the gateway to India, prepared weapons and financial aid for the Afghans after signing an agreement in March 1854. In 1856 Iran decided to capture all of Afghanistan and simply assumed that Britain would cooperate as a result of its 1841 embarrassment in the Khyber Pass. Instead, Britain aided the Afghans. As a result, Iran lost its right of sovereignty in Herat. In November 1856, the British invaded Khark Island, Bushehr and regions from Khozestan to Burazjan. Sir John McNeill, a medical doctor in the British embassy, helped to effect the separation of Herat from Iran (Mojtahed-Zadeh 1998/1377). A month later, Khark Island and Muhammarah (Mahmarah, Khorramshahr) were occupied and British troops continued on to Bushehr and Ahwaz, but war was forestalled when Mirza Agha Khan Nouri accepted terms of peace (Paris Treaty) on 4 March 1857, which included fifteen articles between Iran and Britain (Partu 2009/1388). Farrukh Khan Amin al-Mulk signed the treaty on behalf of the Iranian government.

2.4 Consequences of the Paris Treaty: Boundaries

The Paris Treaty afforded Britain's separation of Afghanistan from Iran, in much the same way as Russia did with Caucasia. The most important articles demanded the following: (i) the Iranian evacuation of Herat and Afghanistan and (ii) the formal withdrawal of any Iranian royal claims to the region, leaving these titles solely to Herat and Afghanistan. Furthermore, (iii) Iran was to abandon all intervention in Afghanistan's affairs; (iv) Herat and Afghanistan were decreed independent; (v) Britain offered to mediate in any conflict between Iran, Herat and Afghanistan.

Hence, the British-India campaign in Kabul and some Iranian cities (Khorramshahr and Bushehr) finally led to the political separation of Iran and Afghanistan from their ancient Persian fealty. To maintain their diplomatic hegemony, the British maintained positions of vigilance, watching for signs of infringement in order to arbitrate any regional conflict. The official document was exchanged on 02 May 1857. The treaty's sixth objective paved the way for the eventual delineation of new frontiers along Iran's eastern borders while Khorasan, Sistan and Baluchistan remained under British control.

After successive wars between Iran and Afghanistan, Britain deemed it necessary to redefine Iran's Eastern frontiers. In 1857, Britain compelled Iran to remove its troops from Herat and obtained an official withdrawal of the Persian Shah's sovereign claim to Afghanistan. Iran also promised to refer all disputes with Afghanistan to British arbitration and judgment. Although the Paris Treaty declared Herat an independent entity, it effectively became another British colony, even though its Shah, Ahmad Shah Baluch, paid taxes to the Royal Court of Persia for another six years after its independence. The dependency of Herat on Iran continued until 1862, when Dost Muhammad Khan, the Emir of Kabul, victoriously besieged Herat and merged the governments of Kabul and Kandahar, eventually leading to the establishment of Afghanistan (Mojtahed-Zadeh 1998/1377).

Iran then accepted the Hirmand River as its frontier border between both countries, but the determination of southern borders, including Baluchistan, was complicated and took another three decades. In 1871, Major General Goldsmith was appointed to demarcate that area. Although his border arbitration commission was merely charged to investigate and consult with an Iranian delegation before the official declaration of their findings in Tehran, the British-India government had given detailed instructions to Goldsmith for a unilateral determination (Mojtahed-Zadeh 1998/1377). The vast area was lost to Britain by the First Goldsmith Arbitration in 1863 and the Second Goldsmith Arbitration in 1872. In addition, a natural disaster decimated the Hirmand River Basin and led to its gradual obliteration, which brought colossal environmental and economic damage to the region, including Sistan, Iran and Afghanistan.

Much has been written on the demarcation of Iran's eastern borders and the difficult circumstances faced at the time. These include several published works, such as Curzon's (1892) Iranian history, *Persia and the Persian Question* (two volumes). Another was written by Lieutenant Yates (1900), *Khorasan and Sistan*, while Pirouz Mojtahed-Zadeh penned the most treasured work (2004), *Small Players of the Great Game* (Yate 1900). Others include Tate's (1909) *The Frontier of Baluchistan—Travel on the Borders of Persia and Afghanistan*, and his (1911) *The Kingdom of Afghanistan* (Tate 1909, 1911). The most important reflections offered here are that all British machinations wrought an environment of complete insecurity for the common people of these regions, as well as the loss of vast Iranian territories, some of which remain in unresolved dispute to this day.

2.5 Conclusion

Following events of the early nineteenth century as the result of cunning British influence in the Royal Court of Fath-'Ali Shah Qajar, treaties were signed that mentioned the name, 'Afghanistan', for the very first time in official documents. At the time, Iranians assumed 'Afghanistan' was a term describing the governor of Afghani tribes. However, the British imputed it to political geography for the purpose of gaining territorial independence from Iran. These treaties were evidently obtained to protect British-India and gain political influence and power in Central Asia. Also, shrewd strategies were employed to influence the Iranian Royal Court, not out of any fear of Iran, but rather to simply gather intelligence on the affairs of Persia's governance. British-India desired that Iran, without fully comprehending why, would not entertain obligations or treaties with Russia, France or any other European nation.

Furthermore, British diplomats endeavoured to keep Iran embroiled in difficulties and economically and politically weakened. In their view, a feeble Iran best guaranteed the continuance of British hegemonic expansion in the region. Furthermore, they joined this strategy with the pretence of perpetual friendship while, at the same time, preferring an unremitting Russo-Persian war. With the prospect of ongoing regional disorder, and although Britain suffered a series of heavy losses in Afghanistan, it survived well enough to achieve its main purpose, which was the separation of Herat from Iran, after which British-India simply delineated new borders.

The Eastern regions of Iran witnessed powerful political pressure from nineteenth-century British-India that led to Iran's loss of treasured territories in the region. Nevertheless, the greatest reason for this loss can be attributed to the incompetence and ignorance of the contemporary local governors and Iranian leadership. They were not only unaware of the actual geopolitical situations regionally and trans-regionally, but also lacked the ability to respond reasonably or impose more favourable treaty obligations. This inherently divisive association of emirs, shahs, sultans and governors also lacked knowledge regarding the legal principles of border partitioning. Hence, they forsook their responsibilities as leaders and lost what God had given them to protect for the want of knowledge and political unity, making them easy prey for the British Lion.

References

Aitchison, C. (1876). *A collection of treaties, engagements, and sunnuds* (Vol. 7). Calcutta: Foreign Office Press.

Bahmani Qajar, M. A. (2007/1385). Vapasin Talashha Baraye Jilugiri az Judaei-e Siyasi-e Iran va Afghanistan (The latest efforts to prevent political separation of Iran and Afghanistan). *Mahnameh Itila'at-e Siyasi Ightisadi (Monthly Politics-Economy Information), 233–234*, 146–157.

Curzon, L. G. N. (1892). *Persia and the Persian question*. London: Longmans, Green, and Co.

Ghani, C. (2013). *Iran and the West: A critical bibliography*. London: Routledge.

Hessam-Moezzi, N. (1987/1366). *Tarikh-e Ravabit-e Siyasi-e Iran va Dunya (History of political relations of Iran with the world)*. Tehran: 'Ilm Publication.

Kaye, J. W. (1874). *History of the war in Afghanistan* (3rd ed., Vol. 2). London: Wm. H. Allen & Co.

Mahmoud, M. (1982/1361). *Tarikh-e Ravabit-e Siyasi-e Iran va Inglis dar Qarn-e Nuzdahum (The history of political relations of Iran and Britain in 19th century)* (4th ed., Vol. 1–3). Tehran: Eqbal Publications.

Mojtahed-Zadeh, P. (1996). *The Amirs of the Borderlands and Eastern Iranian Borders*. London: Urosevic Research Foundation.

Mojtahed-Zadeh, P. (1998/1377). Nigahi be Chigunigi-e Shikl-ghiri-e Marzhay-e Khavari-e Iran (A look to forming of Iran's Eastern Boundaries). *Mahnameh Itila'at-e Siyasi Ightisadi (Monthly Politics-Economy Information), 131–132*, 4–23.

Mojtahed-Zadeh, P. (2004). *Small players of the Great Game*. London—New York: Routledge, Curzon.

Partu, A. (2009/1388). Peiman-e Paris, Dastan-e Ghamangiz-e Judaei-e Herat az Iran. *Itila'at-e Siyasi-Ightesadi (Monthly Politics-Economy Information), 255–256*, 14–41.

Rayfield, D. (2013). *Edge of empires: A history of Georgia*. London: Reaktion books.

Tate, G. P. (1909). *The frontiers of Baluchistan—Travels on the borders of Persia and Afghanistan*. London: Witherby & Co.

Tate, G. P. (1911). *The Kingdom of Afghanistan*. London: Witherby & Co.

Wright, D. (1977). *The English amongst the Persians: During the Qajar period 1787–1921*. London: William Heinemann Ltd.

Yate, C. E. (1900). *Khurasan and sistan*. Edinburgh and London: William Blackwood & Son.

Mansoureh Ebrahimi (Iran/Malaysia) Ph.D., is a senior lecturer with the Faculty of Islamic Civilization, Universiti Teknologi Malaysia. Dr. Ebrahimi has written about Iran's culture, history, religious beliefs and Malaysia's *halal* industry from the Islamic perspective. She also teaches Persian and Philosophy of Science and Civilization. Her books include *The British Role in Iranian Domestic Politics (1951–1953)* (Germany: Springer Nature, 2016); Mansoureh Ebrahimi and Kamaruzaman Yusoff (Eds.): *The Halal Industry in Malaysia: Prospects and Challenges* (Malaysia: UTM Press, 2015); has also contributed several articles on culture, religion and history (including Malaysia, Middle East, Iran and Turkey) to reputable peer-reviewed journals; 'The British plot to remove Iran's democratically elected Prime Minister from power', in: *Acta Histriae* (Republic of Slovenia: Annales Publishing House), 23, (2015): 735–748.

Kamaruzaman Yusoff (Malaysia) is professor at the Faculty of Islamic Civilization, Universiti Teknologi Malaysia, Head of West Asia Civilization Research Group, where he specialises in Middle Eastern History and Islamic Studies. He holds a B.A. in History from the National University of Malaysia; and an M.Litt. and a D.Litt. from the Department of Islamic and Middle Eastern Studies, Edinburgh University, Scotland. He is the author of several articles and books on the Middle East, Orientalism in Islamic History, and the relationship between the Ottoman Empire and the Malay World; and has served as Head of Department Research Group and Dean of the Faculty.

Chapter 3
Origins of Conflicts Between Iran and Iraq

Alireza Najafinejad and Saeid Bahrami

Abstract The Iran-Iraq conflict over a common border became yet another major problem as the result of endless interpositions and transnational intrusions that have caused ceaseless conflicts with long-term consequences in the Middle East. This chapter reveals why mutual agreements between Iran and Iraq have not led to political stability and regional security. The writers review the border conflict's history and causes as well as mutually signed accords between both nations. The investigation methodically uncovers factors that have, directly or indirectly, intensified regional war(s) and ongoing discord.

Keywords Iran's western borders · External influence
Arvand River (Shatt al-Arab) · Ottoman Empire · Oil resources

3.1 Introduction

Iran's unique geopolitical location has always been a place of contention. It is a place where the world's superpowers have routinely played out militant geopolitical strategies as its importance invites their interference and continually motivates their policies of expansion that allow access to Iran's God-given wealth. Special geopolitical factors include enormous oil and gas reserves that supply a major portion of the world's fuel. In addition, Iran's position in the Middle East permits ready transit of goods from Asia to Europe via large ports with strategic access to the Strait of Hormuz and the Sea of Oman. Covetous colonial powers and neighbouring countries

---

A. Najafinejad (✉)
Department of Political Science, Faculty of Humanities, Golestan University, Shahid Beheshti St., 4913815759 Gorgan, Iran
e-mail: ar.najafinezhad@gmail.com

S. Bahrami
Department of Political Science, Faculty of Literature and Humanities, University of Guilan, Rasht, Iran
e-mail: saeed92bahrami@gmail.com

M. Ebrahimi et al. (eds.), *The Dynamics of Iranian Borders*,
https://doi.org/10.1007/978-3-319-89836-0_3

have instigated numerous disagreements and tense differences with the Ottomans and border disputes with Iraq, especially the well-known eight-year war. Nevertheless, determining and protecting geographic borders between Iraq and Iran is an ancient contention that preceded the Ottoman era and recently, with foreign support, led to a war whose destructive consequences still loom large in the region. This chapter examines historical factors and causes of the re-emergence of old tensions along Iran's western front, which in the course of history has been the most sensitive and troublesome of Iranian borders. The author intends to shed light on superpower roles in this contentious region and expose their hidden impositions.

3.2 Circumstances Behind the Emergence of Disagreements Between Iran and the Ottomans

The establishment of the Safavid state in the early sixteenth century, along with the proclamation of Shi'ism as the official religion of Iran, outraged the Sunni Ottoman government, which then adopted a hostile policy (Parsadust 1991/1370: 28). Sunni governments, such as the Ottoman Empire in the west and the Uzbeks in north-east Iran, enjoyed considerable military power. Sunni states and tribes beyond its borders encircled Iran, and within its borders, numerous Sunnis held crucial political significance.

Shah Ismail announced that Shi'ism was the official religion of the country in Tabriz, where the majority of the people there and in Ardebil and Azerbaijan were Sunnis. Shah Ismail established national unity in Iran after a long period of disturbances and disunity. To protect Iran's independence, he used religion as a nationalist umbrella to achieve political independence and Iranian unity. Shah Ismail recaptured various parts of Iran and then invaded the Mesopotamian region (modern Iraq) and occupied Baghdad on 16 October 1508 CE. He destroyed a number of Sunni holy sites in Mesopotamia, killed some Sunni clergymen, and renovated the Shi'a mausoleums of their Imams in Najaf and Karbala (Agoston and Masters 2009: 71). Later, he attacked the north and occupied Mosul in the Kurdish territory of northern Mesopotamia. Effectively, until 1510, Shah Ismail occupied the entire territory of what constitutes modern Iraq. From 1510 to 1555, there were ongoing border conflicts with bloody battles. Eventually, a peace treaty between Iran and the Ottoman Empire was signed after negotiations in the city of Amasieh (Asia Minor) on 29 May 1555. The Amasieh Treaty was the first such treaty between Iranian and Ottoman states, and ended an infamous twenty-year period of war. According to the Treaty, Azerbaijan, eastern Armenia and eastern Georgia were ceded to Iran. Western Georgia and Armenia, as far as Shatt al-Arab (the western Zagros Mountains border), were handed to the Ottomans (McLachlan 1989: 401). The city of Ghares was declared neutral. This treaty of exchange also marked a contract with the Ottoman Sultan, who agreed to treat Shi'ites with tolerance and support Iranian pilgrims to Mecca and Medina. Orders were issued to border commanders to refrain from issuing any commands that might lead to a border dispute (Parsadust 1991/1370: 28).

Since the Amasieh Treaty, until the Algiers Agreement (1975) between the Shah of Iran and Saddam Hussein (mediated by Algerian president Boumedienna), eighteen border dispute resolution treaties were made between Iran and its western neighbour over the course of 420 years. The number of treaties and the duration of disagreements indicate deeply rooted enmity between Iran and its western neighbours, especially in terms of border disputes. They stand as testimony that none of these successfully secured provisions made by either side. Generally, they were imposed on unequal terms and as soon as the discontented party gained greater power, war was renewed. Hence, before the establishment of European influence, especially Britain's hand in the Middle East and Persian Gulf, religion played a crucial role in perpetuating disagreements and frequent wars between the Ottomans and Persia. In addition to sectarian differences (Sunni versus Shi'ite) and respective religious fanaticism, Iranian holy cities in Iraq, including Karbala, Najaf, Kadhimiya and Samaria, remained potent motivating foci of recurrent wars. Ottoman government officials generally mistreated Iranian pilgrims to these sites (Williamson 2008: 89). In response, Iranians desired freedom from oppression when visiting traditional holy cities, which they considered to be within ancient Iranian borders (Barthold 1984: 206). They simply wanted to visit the mausoleums of their revered Imams when they wished and with less expense.

After many wars between the Ottoman Empire and Iran, the Treaty of Zuhab in 1639 was the very first attempt to determine clear borderlines (Schofield 2008: 105). Sultan Murad IV recaptured Baghdad and eventually signed a treaty between both nations in Zuhab. However, this treaty did not determine specific boundaries. Rather, it merely mentioned which areas belonged to which side. This chequered border stretched from the Persian Gulf to Azerbaijan, Armenia and the western side of the Zagros Mountains and Tigris and Euphrates Rivers, and finally, to Basra (Hatef 1992/1371: 48–49; Williamson 2008: 88). The treaty forever ceded Iraq to the Ottomans. Hence, from the three years of 1776 to 1779 when Sadegh Khan (Karim Khan Zand's brother) ruled Basra, the aforementioned territories remained under the dominion of the Ottomans until WWI. The Zuhab Treaty, when viewed as the first such borderline settlement, therefore acquires much significance. Even so, borders were never definitively determined. For example, a strip of land over which neither the Ottomans nor Iran wielded power, was termed the border. As neither country had reliable maps or modern means of transport and communications, they could not exercise clear political control.

Conflicts between Iran and the Ottoman Empire continued after the collapse of the Safavid Empire. Nader Shah, for example, in 1743 and later, fought against the Ottomans but neither side was victorious and measures were taken to establish peace after negotiations leading to the Treaty of Kerden (September 1746) (Parsadust 1991/1370: 65). This pact comprised essence, conditions, terms, and appendices. In terms of essence, both parties agreed to respect the chequered borderlines defined in the Treaty of Zuhab (King Safi and Sultan Murad IV, 1639), the significance being that, for 232 years (since the Battle of Chaldiran in 1514), the Ottomans officially acknowledged that Shi'ite Iranians were Muslims and were to treat them as such, without bias (Hatef 1992/1371: 66–67). Despite its peaceful intent, the Kerden Treaty also failed because ancient religious factors and expansionist spirits were soon disturbed even more by colonialist objectives on the parts of Russia and Britain. The latter nations added fuel that allowed the resumption of even greater armed conflict between Iran and the Ottoman Empire.

3.3 British and Russian Influence on Iran-Ottoman Relations

British sway over Iran took hold during the reign of Shah Abbas, the powerful Safavid Shah. He intended to drive the Portuguese from Hormuz but lacked a naval force. Hence, he inveigled representatives of the East India Company's trading agents in Iran by providing them with plenty of facilities and profitable opportunities. The company acceded to the Persian Shah's demand for warships to confront the Portuguese. Thereupon, with the help of the East India Company's fleet, Shah Abbas drove the Portuguese from the island in 1622 and established dominion. Thereafter, the British position was enhanced despite serious challenges that remained from the Netherlands. Nevertheless, Shah Abbas brought a balance of power between both parties and once again permitted Portuguese castles and trade centres in Kong Port (north-east of Bandar Lengeh). There, the Portuguese fished for pearls while the British gained currency in Iran and the Persian Gulf. The British then adopted a policy that would allow serious impositions in Iran's internal affairs. The Safavid monarchy weakened after Shah Abbas and, along with Afghan aggression, Iranian affairs were soon cast into disarray yet again. This provided a convenient opportunity for the British to secure a better footing. With the establishment of its sway in Iran, Britain considered the Persian Gulf a priority region of importance for trade with its Asian colonies and a fulcrum for manipulating the Ottomans. By 1764, Britain obtained an agreement with the Ottomans to establish a consulate in Basra. By 1798, Britain established an embassy in Baghdad and began positioning its army along the Shatt al-Arab coastline and Kuwait to protect sea-lanes in the Persian Gulf and Shatt al-Arab. An edict issued by the Ottoman Sultan on 29 December 1834 granted Britain the exclusive privilege of navigating the Tigris and Euphrates (Parsadust 1991/1370: 77). Meanwhile, for mutual benefits gained by hampering the French military advance, relations between Britain and Tsarist Russia spawned aspirations of world conquest in opposition to Napoleon's rise (1779 and 1815).

By ratifying the Treaty of Turkmenchay (1828), the Russians gained specific concessions that enabled interference in Iran's internal affairs. Concurrently, the British government was intent on controlling the Persian Gulf and considered Russian access to these waters a threat to its regional interests, especially India. To forestall Russian ambitions in the region and without open war against the vast and powerful Russian motherland, the British adopted a policy of political cooperation with the

Russian government. Furthermore, considering both countries held considerable sway and numerous interests in Iranian and Ottoman royal courts and territories, respectively, they deemed their interests were best served by cooperation (Schofield 2008: 152). This allowed wonderful opportunities for both parties to exploit western Asia freely. Hence, political and military cooperation between Russia and Britain over Iran and the Ottoman Empire was ongoing throughout the entire nineteenth century, through to the beginning of WWI (1914).

Because of oppositional historical and religious backgrounds, and despite a number of treaties, the last of which was ratified in Erzerum in 1823 (the First Treaty of Erzerum), perennial conflicts and hostilities between Iran and the Ottomans hindered both Russian and British governments from enjoying their acquired privileges. Both superpowers deemed the prolonged conflict between Iran and the Ottomans detrimental to all concerned. Therefore, they opted to cooperate and put an end to never-ending border disputes to establish sufficient peace and security for commerce and exploitation of the region's raw materials (Parsadust 1991/1370: 78).

After ratifying the Turkmenchay Treaty, Fath-Ali Shah sent his son, Abbas Mirza (an extremely competent man), to the Azerbaijan borders to put down a rebellion and establish security in Khorasan. Abbas Mirza accomplished the task within two years, which surprised the British, who had always been concerned about its interests in India. Their continued dominion there was threatened by Mirza's success because he might also extract obedience from Afghani opposition after strengthening Iran's power in that region, and then, God forbid, invade India. Another option was closer relations between Iran and Russia after the Turkmenchay Treaty, which could have then paved the way for an incursion of Russian agents into India. The British government, therefore, decided to weaken the Iranian government by inciting Afghan tribal chiefs and thus create new troubles.

Abbas Mirza unexpectedly died in Mashhad (1833) and Mohammad Mirza was elected crown prince. Thereafter the Shah died the following year and Mohammad Mirza, via the resolute stratagem of an astute and knowledgeable minister, Abol-Ghaseme-Farahani, arrived in Tehran and was crowned Mohammad Shah. He was an incompetent and incapable man who murdered his own minister (Farahani) a year after his accession and then appointed Haji Mirza Aghasi, an ignorant man, as his chief minister. The result was great confusion and disturbances and, in some areas, riots broke out. The British government, which favoured the chaos, took advantage to intensify activities in Afghanistan and incited trouble in Herat. Mohammad Shah assembled an army and besieged Herat in 1837. Despite repeated assaults, Afghan forces stood their ground because of British support and military advice. The siege lasted ten months. The British government clearly acted in contravention of an 1814 Treaty with Iran, which stipulated that in the event of war or conflict between Iran and Afghanistan, Britain must not interfere or provide any type of assistance without the consent of both parties. Clearly, against its previous commitment, the British government's support of the Afghans prevented Iran's dominion of Herat. Seeing the plight of the besieged city, Britain then sent warships to the Persian Gulf and occupied Khark Island (Parsadust 1991/1370: 78–80).

3.4 The Second Erzerum Treaty

In 1837, Alireza Pasha, the ruler of Baghdad, attacked the port city of Mahmarah, plundered it, set it on fire, and then reduced it to rubble. British agents undoubtedly incited the action to force Mohammad Shah's assault on Herat. Iran condemned the action, protested against Pasha's invasion, and demanded compensation and appeasement. Instead, the Ottomans responded by saying Mahmarah, a port city, was part of Basra and that Baghdad was theirs as well, therefore only they were justified in punishing their subjects. Hence, if Iran desired appeasement and compensation, Iran must first prove that Mahmarah belonged to Iran. The response added fuel to unfriendly relations, whereupon Iran invaded Sulaymaniyah Province in revenge (1840) and captured it. War seemed inevitable. Russia and Britain, concerned over their own respective interests, especially due to unsecured border disputes, used their influence to mediate and persuaded the Ottomans to relinquish their claims. A joint commission of Iranian and Turkish delegates (with complete authority), as well as Russian and British representatives (mediators), was formed in 1843 and took on the job of determining the borders. Both sides signed a series of negotiations over four years, and on 31 May 1847 a treaty including nine articles was approved. This pact came to be known as the Erzerum Treaty (the Second Treaty of Erzerum). A second treaty of Erzerum addressed all areas with a view to propose solutions (Masters 1991: 9–10). Major disagreements between Iran and the Ottomans concerned (i) indefinite borderlines (both land and sea); (ii) the situation of Kurdistan, especially concerning nomads residing in border areas who constantly moved and whose nationality was uncertain; (iii) the pilgrims who always complained of Ottoman mistreatment and oppression; (iv) Iranian merchants and businessmen who were constantly subjected to the injustices offered by Ottoman customs agents; (v) and the matter of Iranian and Ottoman refugees. Thus, the second Erzerum Treaty was the most comprehensive agreement signed between Iran and the Ottoman Empire (Parsadust 1991/1370: 82).

Iran refused to ratify the treaty and demanded 2.5 million Toman as compensation for the destruction and plundering of Mahmarah. The Ottomans refused. However, Russian and British governments used their influence in the Iranian royal court to persuade Iran to sign. This time, the Ottomans demanded explanations for terms and conditions before final ratification. They sent a letter to the mediating states asking that effects concerning some border issues should be precisely defined. Iran's delegate returned to Tehran seeking authorization to sign the treaty. In his absence, British and Russian ambassadors, without consulting the Iranian delegation, interpreted the treaty's terms as they pleased and officially sent them to the Ottomans without informing the Iranian delegation of the document (Hatef 1992/1371: 52; Schofield 2008: 171–172).

In Article 1, both sides waived financial claims. In Article 2, disputes over land and sea territories were put forward for the first time in the history of treaties between Iran and its western neighbour. Here, the subject of Shatt al-Arab was broached as Article 3 ensured that the settling of all territorial disputes would be well defined, whereupon both sides agreed to the second stipulation and agreed to drop all territorial claims.

Here is where both nations promptly designated representatives to specify exact boundaries according to what was already agreed. In Article 5 both sides considered exemptions of previously agreed rules governing refugees from prior treaties. Article 7 concerned pilgrims and the most common conditions contained in all prior treaties with the Ottomans.

3.5 The 1911 and 1913 Protocols: British and Russian Roles in Their Composition

For forty years, little diplomatic activity ensued over the issue of borders and there were no conflicts over sea boundaries in the Shatt al-Arab River. However, the years leading to WWI found increasing need for definitive contractual boundaries between Iran and the Ottoman Empire. Throughout the nineteenth century, Britain and Russia made serious advances to strengthen and expand their respective positions while increasing respective influences in Iranian and Ottoman royal courts. Developments in the late nineteenth and early twentieth centuries brought them closer and gave rise to increasing interests in the Middle East, especially in Iran and the Persian Gulf. The British wanted to protect trade and communications with India, but this was impossible without a chain of protectorates along the way (Silverfarb and Khadduri 1986: 3). The Russians, who had two wars with Iran, had completely occupied the Caucasus, Yerevan and Georgia, and, on the eastern front, they had captured Turkmenistan and Tajikistan. Detaching these states from Iran was actually part of Russia's age-old March to access warm waters in the south at all costs. Here is where Russian interests met British concerns. The British government desired a protectorate in the south of Iran to support its interests:

> After the First World War, the British government remained in Iraq because it wanted to have military forces near the large British-owned oilfields in south-western Iran and the important oil refinery at Abadan. During the war, this oil had greatly facilitated Britain's military operations, and British leaders believed that in a future conflict it might again become important. Even in peacetime, in the early 1920s, it supplied more than half of the Admiralty's total requirements. However, British leaders feared that if they withdrew from Iraq, Russia would increase its influence in Iran and eventually threaten these oilfields. (Silverfarb and Khadduri 1986: 3–4)

Both nations struggled intensely to gain access to Iran's resources, with each aiding favoured and cooperatively inclined politicians to come to power via incessant conspiracies. They finally divided Iran into two spheres of influence. The north fell to Russia and the south to the protection of Britain. Meanwhile, a third government, Germany, also sought interests in the Middle East, specifically in Iraq, as a market for its products (Silverfarb and Khadduri 1986: 89). Germany had already received the Ottoman franchise to build a railroad from Istanbul to Baghdad and from Baghdad to the Persian Gulf. The Iranian government tried to facilitate a German presence in Iran to escape Russian and British pressure. Hence, when the joint Russo-British commission was busy defining borders, the Iranian government asked the Germans to interfere. Iran's foreign minister and the Ottoman ambassador signed a protocol on 21 December 1911 in Tehran. Both sides agreed to establish a joint commission to settle the border issue. In the event that the commission did not reach a mutually acceptable conclusion within six months, the entire matter was to be delivered to the permanent Hague Court. This Protocol served as the basis for yet another treaty, signed in Istanbul in 1913, and later used as the basis for border redefinitions in 1914 (Hatef 1992/1371: 60–61).

By 1907, Russian and British governments had secretly divided Iran into two spheres of influence cited above. A dividing line that separated both spheres was given to Iran's central government. These powers competed for franchises that licensed (i) the exploitation of mineral resources; (ii) the receipt of national revenues and customs duties; and (iii) monopolies for the import and export of subsequent manufactured products. They encouraged the Qajar Shahs to travel abroad and spend 'borrowed money' in Europe. As the Iranian government was unable to repay these funds, the colonialists demanded more franchises to compensate for incurred 'losses'. Hence, the Iranian people were ever approaching a precipice purposely dug by both states. Their commercial products, especially Russian products, even entered the country without customs duties. Iran's central government in Tehran became powerless to counter these foreign interests or their subjects, or anyone who considered themselves subject to either power.

Meanwhile, negotiations over Iranian-Ottoman border issues pursued spheres of respective Russo-British interests. Russia wished to protect its sphere of influence as delineated in the Erzerum Treaty, while Britain was more than willing to hand parts of Iran to the Ottomans in exchange for lucrative franchises, with Russia's nod of approval. Sir Arnold Wilson, according to his own statements, worked in close cooperation with Sir Percy Cox regarding the final touches of the border issue. This design would hand over the eastern coast up to the riverbank, with the exceptions of Abadan and Khorramshahr, to the Ottomans (Wilson 1932: 270). This confirmed British interests in two areas: (1) the region around Khanaqin where oil wells were part of franchised D'arcy Company land, where boundaries remained undefined; (2) and Shatt al-Arab (Wilson 1932: 215). Hence, Britain pressured Iran to accept these concessions. Ultimately, representatives from four nations, Britain, Russia, Iran, and the Ottoman Empire, ratified the Istanbul Protocol. Although no mention of Russian and British representatives were included in the Tehran Protocol (1911), their agents did not refrain from pressuring both sides of the conflict concerning the matter of borders (Hatef 1992/1371: 62–63).

According to the treaty's first stipulation, for the first time the entire Shatt al-Arab, from the estuary of the River in the Persian Gulf to the Khayyen River (~90 km), came under Ottoman rule. Iran was to be deprived of its undisputed, historical right to common rule over the Shatt al-Arab. The handing over of the Shatt al-Arab to Ottoman rule by Britain and Russia only added more fuel to the fires of disagreement between Iran and the Ottomans. After the Ottoman collapse and the formation of modern Iraq, its ominously detrimental effects dogged Iran and Iraq with a hostile atmosphere for years to come. Historical records indicate that rule over the Shatt al-Arab had never been in the absolute control of the Ottoman government since both countries had always jointly ruled the common border. Sir Arnold Wilson clearly states in his book, *Persia*, that since steamships had entered the Shatt al-Arab, and perhaps much earlier, Iran and the Ottomans had jointly administered all dominion. Now, both sides of the river belonged to the Ottomans. However, the central line of the river, despite the 1847 treaty, and according to Sir Arnold, was deemed the boundary (Wilson 1932: 124). He reiterates in another book, *SW. Persia*, that a decision on the Shatt al-Arab's border was of great significance. For more than fifty years, the borderline was in the middle of the river's main channel or thalweg. By convention, although not precisely stated in the Erzerum Treaty, it had become a tacit agreement (Wilson 1941: 271).

The seventh stipulation of the 1913 Protocol concerned the convention of 28 May 1901, between the Iranian monarchy and William Knox D'arcy (Ebrahimi et al. 2012). Accordingly, these franchises were executable, with all rights and executive authority throughout the entire extent of lands partially ceded to the Ottomans. The 1901 franchise granted to William Knox D'arcy included all contiguous territory in Mesopotamia, from the Persian Gulf to Khanaqin. In the meantime, the British government had become the main shareholder of the D'arcy Company's oil concern (Ebrahimi 2016). Hence, before handing over Iran's territory to Ottoman rule, Britain demanded that the Ottomans recognise this right of franchise. These lands were also near the Qasr-e Shirin oil wells, which promised rich oil reserves.

According to the 1913 Protocol, Britain would enjoy the same franchised rights in lands handed to the Ottomans as if the company still operated in Iran. Available official archives of the Eastern Indian Company and British Foreign Ministry for the previous 300 years by no mean prove Ottoman claims to or prior occupation with long-term possession of Shatt al-Arab. Similarly, the Erzerum Treaty bears no clear terms concerning Ottoman possession of Shatt al-Arab (Hatef 1992/1371: 64). Meanwhile, the development of Abadan's port and increased transatlantic shipping, along with the deepening of the Suez Canal, prompted Britain to ask the Ottomans to deepen Shatt al-Arab. Negotiations were under way, but disagreements once again arose between Iran and Istanbul, not least of which were British interests.

During the course of the Istanbul Commission sessions, Ottoman representatives did not abandon an appendix to the Erzerum Treaty, signed by Mohammad Ali Khan. The Iranian board insisted the document was false and thus invalid, by considering that Mohammad Ali Khan had overstepped his authority. Under pressure from mediating states, the Iranians acceded to the interpretative document (Hatef 1992/1371: 62–63). The 1913 Istanbul Protocol and minutes from meetings concerning border definitions were again discussed in 1914, especially concerning portions of Iran's sea and land territories relinquished to Ottoman control. The legislative bodies of neither country approved these latter details. According to Article 22 of the Iranian Constitution (1907), every law concerning the alteration of Iran's borders had to approved by the Iranian parliament. According to the 1913 Protocol and minutes from meetings in 1914, the entire Shatt al-Arab River, excepting a small patch of 6.2 km facing Khorramshahr, had been relinquished to the Ottoman government. In addition, a 200-mile square parcel of Persian territory around Khanaqin, which later proved rich in oil, was also relinquished to the Ottomans. While this appendix was added to the

Protocol, Ottoman Commissioners refused to attend sessions concerning northern borders, which, on the other hand, they also refused to cede. These regions included numerous territories they had occupied, and which, according to surveyed markings made in 1914, they were meant to relinquish to Iranian control. Moreover, after the joint commission finished, the Ottoman army, while still involved in war against Iran, occupied them in part. The Ottomans, due to their war against Russia and Britain, never approved the 1914 Protocol or minutes.

3.6 History of Disagreements Between Iran and Iraq Over the Shatt al-Arab River and Border

Shatt al-Arab is a border river covering 204 km between Iran and Iraq, situated between 47.5° longitude and 31° latitude, north-west to south-east. A small part of the river comprises the Tigris and Euphrates Rivers, with the remaining portion split between waters from both rivers as well as the Karun River. Shatt al-Arab flows in south-east Iraq just south-west of Iran and empties into the Persian Gulf. Headwaters of both the Tigris and Euphrates are in the Turkish mountains. In addition, several rivers with headwaters in Iranian mountains flow from Iran into the Tigris. The most important of these rivers are Great Zab, Little Zab and Diyala. However, the most vital waters are those of the Karun River, which contains the largest volume of water of all the rivers feeding Shatt al-Arab. The headwaters of the Karun River are in the Bakhtiari Mountains and flow through Iran to Shatt el-Arab near Khorramshahr. Iran has two large seaports on the Shatt al-Arab. Abadan is the largest, mostly for the export of oil products. Khorramshahr is also an important port of foreign trade in Iran. Another important port near Shatt al-Arab is Khosro Abad, which, prior to WWII, was dedicated to oil exports. One of the important Iraqi ports on the Shatt al-Arab is Basra, chiefly for foreign trade. Another is Fav Port, mostly used for the export of oil products.

WWI brought sweeping changes to the region:

> The Sykes-Picot Agreement of 1916 awarded Britain direct control over Basra and Baghdad, as well as a 'zone of influence' over most of the rest of southern and central Iraq. The status of the northern province of Mosul remained contested until after the war. (Agoston and Masters 2009: 283)

The immediate effect of the war on Shatt al-Arab was the establishment of a port office and maintenance management under the complete control of the British government in 1916. Before the war, large transatlantic ships used Iranian guides from Khark Island to navigate the Shatt al-Arab. The war increased traffic between Basra, Khorramshahr, and Abadan, and from this new beginning, the right coast of Shatt al-Arab was occupied by British and Indian forces, and the left coast was also under British occupation. Hence, there was a need for extra guides who required training. This organization was responsible for all matters concerning river and canal maintenance, thus putting an end to the situation whereby criteria for action were based on the river's median and shared responsibility between Iran and the Ottomans. The area which had come under British supervision included waters belonging solely to Iran, a neutral country during the war. The effect thus nullified Iran's basic rights outlined in the 1913 Protocol (Hatef 1992/1371: 69–70). After the war, former powers previously interested in border issues were no longer there. The Ottoman Empire fell and Russia was indisposed with its Bolshevik migraine. Because of expanded Iranian oil production and economic development of an area previously under Ottoman auspices, a considerable increase in river traffic took place. Hence, after the war, the Mesopotamian Region was called 'Iraq' and became The League of Nations protectorate under British supervision.

3.7 The 1937 Treaty and Britain's Intervention on Behalf of Iraq: Reasons and Methods

WWI victors, mainly Britain and France, realigned vast territories of the Ottoman Empire in Europe and western Asia in accordance with the Sèvres Treaty (10 August 1920). They essentially established new countries, which, in practice, came under the protection and influence of the conquering states.

With the establishment of Iraq, the British, with the approval of The League of Nations, supervised its affairs under the title of 'Protector'. The British government was eager for Iran to recognise Iraqi independence so that Iraq would then become The League of Nations member. However, Iran's government had long been in fundamental disagreement with the Ottoman Empire over what now became Iraq. Hence, the matter of Shatt al-Arab made recognition of Iraq's independence dependent on the latter's agreement to recognise Iran's indisputable rights to Shatt al-Arab while also respecting the rights of Iranian nationals in Iraq who were under persecution (Potter 2004: 67). Persia also took the position that it was not bound by treaties negotiated by governments no longer in power, which included both the Ottomans in Iraq and the defunct Qajar dynasty (Sevian 1968: 219).

Because of Iran's insistence, the British plenipotentiary in Iran sent a document to the Iranian Foreign Minister on 11 March 1929 regarding Iranian demands:

> I have just received orders to assure you that if the Iranian government recognises Iraq's independence, my government, after being informed about the Iranian government's opinions on the issue of the practical problems resulting from the current situation of Shatt al-Arab, and the guarantees demanded by the Iranian government, will mediate with the Iraqi government and will help Iran to achieve its reasonable demands. (Iran Ministry of Foreign Affairs 1969: 11)

The key cause of disagreement was the boundary along the Shatt al-Arab, to be discussed below. However, the British also maintained that the frontier had been settled and they blamed Iran for strained relations during the post-war period (Potter 2004: 68).

After official British assurances, and after the Iraqi government lifted some restrictions on Iranian nationals in Iraq (Potter 2004: 69), the Iranian government recognised Iraq's independence, after which the first Iranian embassy opened in 1929. Nevertheless, Iraqi and British governments, despite promising to fulfil reasonable demands, in practice not only did nothing to respect Iranian rights but also showed that such promises were made only to procure Iranian recognition of the newly established government of Iraq. After realizing its mistake, in document number 3276 (1 December 1932), the Iranian government declared the borders of 1914 invalid and refused to recognise them. Thus, began yet a new border conflict, added to already strained relations (Parsadust 1991/1370: 163–164).

King Faisal I demonstrated goodwill towards resolving conflicts between Iran and Iraq. He arrived in Iran on 26 April 1932 to win approval by adopting a friendly approach to negotiations. Iran demanded that Iraq recognise the Shatt al-Arab Thalweg as the borderline (Potter 2004: 72). After preliminary negotiations, Reza Shah, Shah of Iran decided to visit Iraq the following spring to resume negotiations. However, King Faisal died in September of 1933 and the trip did not materialise. Intense disagreements ensued and the border dispute escalated. Iraq insisted on the Erzerum Treaty and the '1914 minutes' on the border issue. However, Iran believed that the treaty had been imposed, and therefore considered it worthless. Meanwhile, Iraq used contemporary border incidents and formally informed the Iranian government of circumstances while relying on Article 2 of The League of Nations Convention. Iraq raised a complaint and informed The League of Nations of an impending risk of war between Iran and Iraq over the borders. Iraq petitioned The League of Nations and mentioned the Erzerum Treaty, the 1913 Istanbul Protocols and the '1914 minutes', stating that all were ratified via the mediation of Russia and Britain. Iran's response to The League of Nations (8 Jan 1935) revisited the history of the Istanbul Erzerum Treaty and concluded that Iran's delegate had exceeded his authority in approving the treaty. Therefore, the treaty was neither binding nor brought obligations to Iran. Regarding the Istanbul Protocol, Iran argued that neither nation's legislative assemblies had approved either the Protocols or the '1914 minutes'. Hence it posited that a protocol designed to confirm the terms of the Erzerum Treaty was equivalent to the treaty and was therefore also invalid (unlawful) (Hatef 1992/1371: 74–75). Iran definitively concluded that claims and arguments made by the newly fabricated Iraqi government were based on invalid documents and thus baseless. It further argued that because no valid documents defining the Shatt al-Arab border existed, the natural border according to conventional international regulations was the thalweg line.

This issue was sent to the International Council (26 May 1935) for both sides to express their opinions. The Soviet Union's delegate, as successor to Tsarist Russia as one of the countries involved in the Istanbul Protocol negotiations, denied any responsibility for the Protocol. He further affirmed that Tsarist officials signed the document and that the current Soviet government did not share common policies or interests with the former regime. The representative of the Turkish government, as chairperson of the session, followed suit and endorsed the Soviet representative's position. He further denied Turkish responsibility, arguing that the documents carried signatures from the Ottoman Empire, which did not concern the current Turkish regime at all. The Italian representative, assigned as secretary for the Council, proposed an obvious conclusion, that an acceptable resolution could not be reached and that it would be better to postpone negotiations until the following September, in the hope that during the interval, both sides might find an acceptable solution. He also recommended that both sides not take any measures that might escalate the conflict. Iran and Iraq accepted this proposal (Hatef 1992/1371: 75–76).

The Iraqi Foreign Minister, Justice Minister and the British ruler of Basra arrived in Tehran in August 1935 to hold further negotiations. Noori Pasha, in the presence of Reza Shah, affirmed Iran's recognition of the thalweg line as the border with Abadan (Hatef 1992/1371: 77–78). This agreement then became the basis of the border treaty of 1937. Although the resolution did not protect Iran's interests fully, the Iranian government was not opposed to it. Iranian and Iraqi foreign ministers then sent letters to The League of Nations Secretary General, after which the Iraqi government officially withdrew its complaint in August 1937. Nevertheless, negotiations once again encountered problems.

Noori Said, the Iraqi Foreign Minister, believed the British should be involved in negotiations, but the Iranian board of representatives did not agree. The British had fully intended to take part in the treaty's ratification and even planned to place a representative in the joint commission management office of the Shatt al-Arab River. Obviously, Britain's policy was to maintain control of the Shatt al-Arab by any means possible. The British were allies and supporters of the Iraqi government and hoped to maintain full supervision of the Shatt al-Arab and protect their access in perpetuity. Their intent was to assure that, in the event of any later treaty, the Shatt al-Arab River would remain under Iraqi dominion with rights reserved for total control and supervision. Iraq's reserves would naturally fall under British supervision and remain so as a British ally. In any case, Iraqi and allied ships could traverse Iraqi waters for the entire course of the river as far as Basra without entering Iranian waters (Hatef 1992/1371: 79–80). Therefore, the British government's behaviour concerning Iran's claim to the Shatt al-Arab had changed from its 1912 position, whereupon some Iranian concessions had to be granted. For this reason, the British Foreign Ministry repeatedly stressed that the contemporary situation limited Iran's rule to its own coasts. Hence, governance of the river by Iraq countered international law (Schofield 2008: 154). Initially, the British Foreign Ministry intended to recognise the thalweg line and grant complete control to Iran, but they were forced to retreat after facing intense opposition from the British Naval Ministry. The reason for the latter's opposition was that if Iran succeeded in obtaining joint governance of Shatt al-Arab by signing a treaty with Iraq, then, according to international rights, Iran could stop British warships entering its waters. Therefore, the idea of the thalweg border for the entire course of the Shatt al-Arab River was rejected.

The British government realised that an Iranian harbour extending as far as the thalweg line facing Abadan would prevent any other threats to its own control over Shatt al-Arab. When it was decided that part of the sailable canal be handed to Iran, Britain's primary concern during negotiations was Iranian governance of the waters. The first problem concerned the free passage of British warships under any circumstances, whether war or peace, without damaging Iran's neutrality. The second proposal was even more important as it concerned the maintenance of the waterway. Undoubtedly, if governance of the river was equally distributed, some kind of cooperative joint control of inspection and maintenance management needed to be in place. Even so, Britain not only sought the assurance of permanent navigation, but also wished to put itself in the position of absolute rule over the river. Hence, it wished to interfere in the joint commission's deliberations over river maintenance and impose its opinion on both countries.

Here again, the British Navy exerted intense pressure to maintain its control of the River. The Naval Ministry proposed that an independent joint advisory commission with representatives from Iraq, Iran and Britain be tasked with the responsibility of coordinating maintenance and surveillance activities, thus carrying the authority for urgent and feasible decisions in case of disagreements. The chief problem with such a convention was that it jeopardised the entire treaty. Britain insisted on being an equal partner to any agreement with equal rights, and in some cases, with the decisive power to impose resolutions on both sides. Nevertheless, it became quite clear that Iran would never accept a River Maintenance Convention signed by Britain as a third partner. Nor would Iran establish a Council with any prospect of British government interference. Conversely, the Iraqi government determined not to accept a Maintenance Convention without British representation (Hatef 1992/1371: 87). The quarrel over this issue was so fundamental that the Treaty itself concealed the entire issue and bore no mention of it whatsoever. Hence, the Treaty was written with remarkable ambiguity regarding the subject of signing a Maintenance Convention for the River. Incomplete preparation for such a convention can be cited as the origin of subsequent disagreements.

3.8 The 1937 Border Treaty

In addition to the Border Treaty of 4 July 1937, a protocol was signed whose fifth article became indelible. The Treaty contained six articles. Article 1 stated that the Iranian government, despite all previous objections, accepted the Protocol of 4 November 1913 plus the '1914 minutes' regarding border definitions, and that both sides were bound to respect these provisions. The article further mentioned particulars introduced in Article 2 as exceptions. For the first time in all of Iran's western border treaties, Article 2 cited the thalweg line. However, despite the fact that it was/is the acknowledged international method for determining boundaries for frontier rivers, the Article did not include the entire course of Shatt al-Arab; i.e., from contact with the Karun River to the Persian Gulf. It encompassed only the small extent facing Abadan, which is about four miles, whereupon the Article immediately stressed that the remaining borders were unchanged from those drawn during the 1914 meetings. Article 3 stipulated that border signs would be set and, if necessary, new signs were to be erected according to the determinations of the 1914 border definition commission. Article 4 was of special importance. Clause A decreed that Shatt al-Arab would be equally opened to commercial ships from all countries. All received charges were to be considered remuneration and exclusively spent on ship maintenance, route improvements and sailing expenses as allocated by the government to improve conditions for navigation. Clause B of Article 4 acknowledged that the Iraqi government tacitly relinquished exclusive possession of the Shatt al-Arab. Article 5 concerned both nations and therein confirmed that navigation in the Shatt al-Arab should bear joint interests with strong testimony to this effect. Article 6 contained an exchange of documents that clearly stated the Treaty in Arabic, Persian and French, and that in the case of disagreement, the French text remained the valid legal reference.

When studying the 1937 Border Treaty and appended protocol, the impression is that because Iraq succeeded in obtaining Iran's approval of the 1913 Protocol and October '1914 minutes'—in which part of Iran's territory was ceded to the Ottoman Empire and its successor, namely Iraq, which had hampered Iran's official joint rule of Shatt al-Arab, excepting the small course facing Abadan Island and the Khorramshahr Port, which came under the rule of both countries—the remaining part, that is, the whole surface of the River throughout its border coursing, on which one side was Iranian territory, had been relinquished to Iraq. Therefore, Iraq must have been eager to implement the 1937 Border Treaty. However, the actual course of events indicated the contrary. The Iraqi government did not pursue Article 4 of the Treaty regarding remunerations, expenditures and implementation of Article 4, which therefore necessitated another treaty concerning Article 4. Moreover, despite the fact that according to this treaty Iraq was obligated to submit an account of received remunerations and expenses, they did nothing in this regard until 30 March 1940, and only once informed the Iranian government of such matters, and even then only in brief, inexpressive and inadequate terms. All Iranian efforts to parley Iraq respecting these obligations proved fruitless during the thirty-two years of the treaty's validity (Parsadust 1991/1370: 210–211). Consequently, tensions mounted and were ever increasing. They reached their highest point in the following instances.

After the establishment of Abd al-Karim Qasim's rule in 1958, relations between Iran and Iraq deteriorated and Iraqi officials and newspapers launched a cold war propaganda campaign against Iran. In such an environment, the possibility of reaching an agreement only diminished. The Iranian government, as in the past, contented itself with an exchange of documents and visits. However, for the first time, a document submitted to the Iraqi Embassy in Iran by the Iranian Foreign Ministry (October 1958) mentioned that if the Iraqi government did not elect a board to resolve land border conflicts and affairs concerning Shatt al-Arab by 6 November 1958, the Iranian government reserved the right to take whatever actions it deemed necessary. No response was received from the Iraqi government and another document was subsequently prepared in November the same year.

The Anglo-Iranian Oil Company in Khosro Abad (on the east coast of Shatt al-Arab, Iranian side) had established a port for the export of petroleum products. After nationalization of its oil industry, the Iranian government used Khosro Abad port facilities to position a naval war fleet and coastguard. After the Iran-Pan American Oil Company signed a treaty with Iran for the discovery and production of oil, it demanded use of Khosro Abad Port as a centre for excavations, to which the Iranian government agreed. However, in 1959, twenty-two years after the treaty of 1937, when the first Iran-Pan American ship approached the estuary of the Shatt en route to the Khosro Abad Port, Iraqi authorities stopped it and announced that they did not recognise Khosro Abad as a seaport.

The Iraqi government, taking advantage of impotent and irresolute Iranian reclamation of rights, was not only engaged in guiding ships but also in all affairs of the Abadan Port and Iranian waters, as performed by Iraqi officers from the Basra Port. In 1960, twenty-three years after signing the 1937 Treaty, the Iranian government decided to take charge of port affairs through the Port Authority of Abadan. In August 1960, they informed shipping companies that Iranian staff, opposed by the Basra Port Office, would handle all shipping matters. The Iranian government, as in the past, engaged in an exchange of documents, and in October 1960 they protested against a move made by officials in Basra by sending yet another document. However, because meetings with Iraqi officials were fruitless, Iran decided to act. It declared that from 16 February 1961, all affairs of the Abadan port concerning the berthing of ships would be handled in Abadan. The Iraqi government responded by warning all shipping companies that any ship not using Iraqi staff for port affairs in Abadan would not be provided with guides to cross the Shatt al-Arab and even promised to make the crossing troublesome. Subsequently, Abadan, the largest oil refinery in the world, was practically closed for nine weeks. Iran incurred enormous losses because of the blockade of refined oil exports, and was forced to retreat. Once again, as in the past, Iran surrendered Abadan's port affairs and its sovereignty of Iranian waters to Iraqi staff in Basra (Parsadust 1991/1370: 216–217).

Following the remarks of Abd al-Karim Qasim and the cold war propaganda between both countries, Abbas Aram (Iranian Foreign Minister) delivered a speech to the Iranian National Assembly on 9 December 1959. He repeated previous charges, saying that the Iraqi government had not fulfilled its obligations concerning the implementation of Articles 4 and 5 of the Borderline Treaty, and Article 2 of its Protocol. He went further by stating that since the Treaty's ratification, Iraq assumed charge of the entire bureaucratic administration of the Shatt al-Arab without Iranian interference. He then accused Iraq of spending the waterway's revenues on extrinsic matters in flagrant disregard of the Treaty. Furthermore, he reported that 75% of these revenues came from Iran, whereupon he proposed that the affairs of Shatt al-Arab should fall under Iran's supervision and management in due proportion. In closing, Abbas Aram stressed, as repeatedly pointed out to the Iraqi government, that Iran would stand firm on the issue of Shatt al-Arab and was determined to defend and exercise these rights. Nevertheless, Iran took no effective measures to indicate such resolve (Parsadust 1991/1370: 216). An intense conflict between the Iraqi government and Barezani Kurds 1965 had meanwhile caused Iraq to violate Iranian borders under the pretence of pursuing Soviet forces. The Iranian government warned Iraq that it was running short on tolerance and that patience with peaceful intentions had limitations (Parsadust 1991/1370: 217). The Iraqi Defence Minister, General Hardan Takrini (head of a board comprising the Iraqi Foreign Minister, the political secretary general of the foreign ministry, the Air Force commander and the Iraqi vice president), visited Iran in December 1968. He expressed Iraq's willingness to resolve all border disputes and the matter of Shatt al-Arab. Nevertheless, no acceptable outcome was forthcoming. Therefore, the Iranian government's documentation warriors declared the following: (i) considering statements of the Iraqi representatives and proposal as the draft of yet a new Treaty; (ii) and because Iraq had indisputably abrogated the last Treaty with violations of several important articles for more than ten years; (iii) and further because the current treaty disregarded clear international convention and law; Iran therefore declared the treaty worthless, false and ineffective (Parsadust 1991/1370: 218).

The Iraqi government had always used violence as a policy toward Iran. Hence, they showed a strong hand in response. They invited the Iranian ambassador in Baghdad to their Foreign Ministry office in April of 1969. The Iraqi foreign deputy minister then declared the following: (i) the Iraqi government considered Shatt al-Arab its territory; (ii) he demanded that Iran decree that all ships carrying the Iranian flag in Shatt al-Arab must lower them; (iii) all Iranian Naval personnel on board any such ship must disembark in deference to Iraqi force; (iv) Iraq would no longer permit ships heading to Iranian ports to enter Shatt al-Arab. Following this, Iran realised it was left with no other option than to respond in kind. On 19 April 1969, the Iranian Foreign Deputy Minister, Amir Khosrow Afshar, addressed the Iranian Senate where he reported Iraqi violations and inhumane treatment of Iranian nationals in Iraq. For these reasons, the 1937 Border Treaty was officially abrogated by the Iranian monarchy and considered worthless, false and ineffective. In addition, the Iranian monarchical government did not recognise or accept any principle regarding Shatt al-Arab other than the international convention, namely, the thalweg or central line of the river. Then, the Iranian government officially declared it would prevent any violation of its right to rule over Shatt al-Arab by force and would not tolerate any violations (Parsadust 1991/1370: 225–226).

3.9 The Algiers Agreement and 1975 Border: Good Neighbourly Relations Treaty

From 4–6 March 1975, the annual OPEC Summit was held in Algiers. By then, conflicts between Iran and Iraq had peaked, along with the effective ‘all out’ support of Iran’s Shah for Mustafa Barezani’s Kurds. Iraqi failure to suppress the Kurds caused serious problems for their government. Houari Boumédiène (Hawārī Bū-Madyan*)*, Algiers’s president and host of the OPEC Summit, acted as mediator between Iran’s Shah and Saddam Husain. Afterwards, a joint declaration was issued on 6 March 1975, whereby foreign ministers of Iran, Iraq and Algiers agreed to meet frequently to negotiate resolutions for all contentions. On 13 June 1975 they signed a ‘Border and Good Neighbourly Relations Treaty’ comprising (i) protocols concerning the re-marking of land borders; (ii) protocols concerning river boundaries; and (iii) protocols concerning border security. The 1975 ‘Border and Good Neighbourly Relations Treaty’ was signed in accordance with principles agreed to by both countries, as contained in the Algiers Declaration of 6 March 1975. The Algiers Accord includes an introduction and four principles respecting territorial integrity without violation of borders, and non-interference in each country's internal affairs. Both sides agreed on the following points:

1. To define their borders according to the Constantinople Protocol of 1913 and '1914 minutes';
2. To define sea borders according to the thalweg;
3. To establish peace and mutual trust along the common border.

Both sides agreed that the above content was inseparable from a general solution and that violating any term countered the spirit of the Algiers Accord.

After issuing the joint declaration, the foreign ministers of Iran and Iraq met in Algiers frequently to negotiate further. On 13 June 1975, they signed the Border and Good Neighbourly Relations Treaty in Baghdad. It contained protocols for re-marking land borders, defining river borders, and border security, with an addendum concerning Clause 5 of the Treaty's Article 6. Some notable terms of the Algiers Accord are presented below:

A. Regarding land borders: Article 1 confirms the land borders between both countries as defined according to protocol. Accordingly, the re-marking of land borders was to be done by a joint committee from Iran, Iraq and Algiers according to the 1913 Constantinople Protocol and '1914 minutes'. To define these borders and the river's thalweg, and then erect border signs, technical committees comprising experts from both countries under the supervision of Algiers representatives were sent to Shatt al-Arab prior to closing these borders between both countries.
B. Regarding river boundaries: Article 2's protocol defined the river border between Iran and Iraq as the thalweg, beginning from the Shatt al-Arab's meeting with the Khayyen River, through to the sea. In the case of changes in the main channel due to natural causes, the border was to remain at the thalweg.

This treaty's definition of the common border put an end to one of the oldest conflicts between Iran and Iraq. Nevertheless, the 1975 Treaty makers failed to anticipate the right of unilateral abrogation.

3.10 Reasons for the Abrogation of the Algiers Agreement by Iraq

On 17 September 1980, Saddam Husain unilaterally abrogated the 1975 Treaty. In justification, he accused Iran of intending to separate part of Iraq's territory by providing financial and military support to the Kurds in northern Iraq. He also accused Iran's leaders, since their coming to power, of deliberate interference in Iraq's internal affairs by supporting the Kurds and providing them with arms. He further claimed that Kurds were sponsored by Zionists and America, exactly like the Shah who previously supported them and refused to return Iraqi land. Therefore, the Iraqi government had no choice but to free its lands by force. The Iraqi government declared absolute dominion over the entire Shatt al-Arab and threatened intolerance for any violation of its authority. After the abrogation, Iraq took urgent measures to prohibit activities that had been in place since the Iraq-Iran joint rule over the Shatt al-Arab. To exercise its exclusive rule over the Shatt al-Arab, Iraq declared that all ships must relinquish 'right of passage' exclusively to Iraqi officials and that only Iraqi guides could be used. The Iraqi government formulated new rules to maintain ship security and warned all ships that they must respect these rules, emphasizing that on entering the Shatt al-Arab, they must fly only the Iraqi flag, even if they intend to sail to Iranian ports.

From an international rights perspective, disagreement over implementation of a treaty never justifies unilateral abrogation. Furthermore, according to Clause 3 of Article 62 of the Vienna Convention, even considerable changes in circumstances do not permit unilateral abrogation or withdrawal from a treaty, but can justify a request for the suspension of a treaty's enforcement. Clause 2 of Article 62 of the Vienna Convention clearly states that an argument for 'considerable change in circumstances' as the reason for unilateral abrogation of a Border Treaty is invalid. In addition, the *Introduction* of the UN Charter, the most important international legal and political document, considers the principle of respect for treaties obligatory for its members (Valdani 1998: 48). Generally, governments respect a few major principles concerning bilateral and multilateral treaties. These include respect for international obligations, respect for territorial integrity, and the avoidance of force in lieu of peaceful negotiations. International rights, in light of respect for territorial integrity, refers to components of border treaties as certain, permanent, and unalterable. As a general rule, it is not only unacceptable to unilaterally alter a treaty based on 'considerable changes in circumstance', but also invalid, regardless of any claims, including *ineffectiveness* or *suspension of execution* of terms, or *expiration* of treaty principles. The validity of a treaty, according to conventional 'international rights', reserves alteration of treaties ratified by former colonial governments for the purpose of 'achieving the independence of new countries'. It also prohibits the execution of this principle concerning border treaties by emphasizing the principle of treaty sustainability. This was the significant international principle utilised by Iran and Iraq when they ratified the 1975 Treaty, and is clearly stated in Article 5 of the treaty as follows: "... *concerning the inalterability of the borders and respect for territorial integrity, both sides confirm that their land and river boundaries will be inalterable, permanent, and definite*".

3.11 Reasons for 1975 Treaty Validity: The Iranian Perspective

1. The 1975 Treaty embraced principles of governance via 'international rights' by determining the Shatt al-Arab Thalweg as the border between Iran and Iraq.
2. War does not breach the validity of this treaty. As to whether war justifies abrogation, international rights scholars express two theories: (i) some believe that

with the outbreak of war, treaties lose validity. (ii) However, they exclude border treaties from this ruling and posit that war does not breach the validity of such treaties. Conversely, some believe that many treaties retain validity after the outbreak of a war, when the sole effect is to postpone implementation.

3. Article 4 of the 1975 Treaty states that the treaty and its three protocols and appendices are absolute, permanent, and unbreakable. Thus, the 1975 Treaty was intended to remain valid at all times, including war (Valdani 1998: 52).

The most important and worthy point is that Iraqi government publicly endorsed this issue in 1969 when the Iraqi representative in the UN sent a letter to the Security Council stating the following:

> The unilateral abrogation or altering of the border treaty is invalid under any circumstances, even if there is war between both countries; this is an absolute law. In addition, according to the international principles and rules, the unilateral abrogation of the 1975 Treaty is condemned and this treaty, its three protocols and appendixes, plus four supplementary agreements are valid and will remain so. (Valdani 1998: 52)

References

Agoston, G., & Masters, B. (2009). *Encyclopedia of the Ottoman Empire*. New York: Facts On File, Inc.

Barthold, V. V. (1984). *An historical geography of Iran* (S. Soucek, Trans.). Princeton: Princeton University Press.

Ebrahimi, M. (2016). *The British role in Iranian domestic politics (1951–1953)*. SpringerBriefs in Environment, Security, Development and Peace: Germany.

Ebrahimi, M., Yusoff, K., & Mohamad, A. B. (2012). British responses towards Iran after approving the nationalization of oil in 1951. *International Business management, 6*(2), 277–285.

Hatef, M. (1992/1371). *The relations between Iran and Iraq*. Tehran: Marvi Printing House.

Iran Ministry of Foreign Affairs. (1969). *Some facts concerning the dispute between Iran and Iraq over the Shatt al Arab*. Tehran: Ministry of Foreign Affairs Publication.

Masters, B. (1991). The treaties of Erzurum (1823 and 1848) and the changing status of Iranians in the Ottoman Empire. *Iranian Studies, 24*(1–4), 3–15.

McLachlan, K. (1989). Boundaries with the Ottoman Empire. In E. Yarshater (Ed.), *Encyclopaedia Iranica* (Vol. 4, p. 401). London—New York: Routledge and Kegan Paul.

Parsadust, M. (1991/1370). *Rishehaye Tarikhiye Ekhtelafate Iran va Iraq, 1514–1980 (The Historical roots of the conflicts between Iran and Iraq, 1514–1980)*. Tehran: Sherkate Sahamiye Enteshar.

Potter, L. G. (2004). The evolution of the Iran-Iraq Border. In R. S. Simon & E. H. Tejirian (Eds.), *The creation of Iraq, 1914–1921* (p. 61). New York: Columbia University Press.

Schofield, R. (2008). Narrowing the frontier: Mid-nineteenth century efforts to delimit and map the Perso-Ottoman border. In R. Farmanfarmaian (Ed.), *War and peace in Qajar Persia: Implications past and present* (p. 149). New York: Routledge.

Sevian, V. J. (1968). The evolution of the boundary between Iraq and Iran. In C. A. Fisher (Ed.), *Essays in political geography* (pp. 211–223). London: Methuen.

Silverfarb, D., & Khadduri, M. (1986). *Britain's informal Empire in the Middle East: A case study of Iraq 1929–1941*. Oxford: Oxford University Press.

Valdani, A. J. (1998). Barresi-ye Ahdname-ye 1975, Iran va Iraq (Review of 1975 Iran-Iraq treaty). In *Mahnameh Itila'at-e Siyasi Ightisadi (Monthly Politics-Economy Information), 247–248*: 40–55.

Williamson, G. (2008). The Turko-Persian war of 1821–1823: Winning the war but losing the peace. In R. Farmanfarmaian (Ed.), *War and peace in Qajar Persia: Implications past and present* (p. 88). New York: Routledge.
Wilson, A. (1932). *Persia*. London: Ernest Benn Ltd.
Wilson, A. T. (1941). *SW. Persia: A political officer's diary, 1907–1914*. Oxford: Oxford University Press.

Alireza Najafinejad (Iran) is an Assistant Professor of Political Science at Golestan University (Iran). He holds a B.A. in political science from Mashhad University; an M.A. in Political Science from Karaj Islamic Azad University; and a Ph.D. in Politics and Government from the University Putra Malaysia (UPM), Malaysia. His areas of interest include Middle East Studies and Human Rights in Islam.

Saeid Bahrami (Iran) is an independent researcher who specialises in Iranian foreign policy. He holds a B.A. from Chamran University, and an M.A. in International Relations from the University of Guilan (Iran).

Chapter 4
Evolution of the Post-soviet Caspian Sea Legal Regime

Abdollah Baei Lashaki and Masoumeh Rad Goudarzi

Abstract Despite more than two decades since the collapse of the Soviet Union and several conventions and meetings, Caspian littoral countries have been unable to agree on the Sea's legal regime. Geography, self-allocations of coastal regions, the race to develop massive oil and gas reserves, and the presence of trans-regional powers have all contributed to disparate views on possession and the determination of boundary lines. This chapter reviews successive agreements between Iran and both the Russian Empire and the Soviet Union. This is then followed by a discussion of sundry reasons for the failure of littoral countries to reach a consensus on a new legal regime with a special focus given to political and economic competition between regional and trans-regional players.

Keywords Caspian Sea · Legal regime · Legal status · Caspian Sea littoral states · Energy resources · Caspian Sea cooperation organization

4.1 Introduction

In principle, crises in international relations are over the acquisition of geographical space. Their histories have a relatively stable continuum because values and benefits of such disputes always involve access to privileges and resources that are politically tangible as disputing parties seek to maximise respective interests for as much available space as possible. The legal regime of the Caspian Sea has a similar history, which, especially since the Soviet collapse, has attracted the keen devotion of

A. Baei Lashaki (✉)
Department of Law, Applied Art and Culture Center, Science—Education University, Nowshahr, Iran
e-mail: abdollah.baei@gmail.com

M. Rad Goudarzi
Department of Political Science, Faculty of Literature and Humanities, University of Guilan, Rasht, Iran
e-mail: Fr.raad1@gmail.com

M. Ebrahimi et al. (eds.), *The Dynamics of Iranian Borders*,
https://doi.org/10.1007/978-3-319-89836-0_4

regional politicians. Despite several years filled with conferences, working groups and numerous meetings at different governmental levels of the concerned littoral countries, these nations have yet to reach any agreement that clearly determines the Caspian Sea's boundaries. However, all concerned parties express the desire to do so.

Before the collapse of the Soviet Union (1991), the Caspian Sea was known as the Iranian-Soviet Sea according to a 1921 Treaty, and re-designated as such in 1940 as the joint maritime territory of both countries. In practice, the Soviet Union, as a superpower whose territory surrounded the Caspian Sea on three sides, exerted far more influence on the sea. Nevertheless, no serious legal problems arose between either nation during this period and any lesser issues that did occur were settled on the basis of the 1921 or 1940 treaties through bilateral negotiation. With the collapse of the Soviet Union and the rise of newly independent Caspian states however, the geopolitical situation changed as littoral countries increased from two to five. Hence, attitudes towards Caspian Sea problems also changed along with reconsideration of its legal regime. Then began a series of different approaches involving the exploitation of under-seabed resources. Since then, several discussions/meetings over 'how to determine the legal regime' have been held on the sea's division, none of which has achieved either certainty or unanimity. Each country prefers to consider its own national interests.

Geographical features of this largest lake in the world now embrace five littoral countries, each of which has allocated itself part of the coast and longitude, portions of which hold significant amounts of oil and gas. Trans-regional powers have different views on the Caspian Sea's dominion and determination of its boundaries.

For more than seventy years, approximately 22 million km^2 of the region was under socialist rule, through which the communist system comprehensively monitored all aspects of societies within its territory. There was no possibility or question of external influence or impositions during this time and the Western world never got close to the vicinity. However, the collapse of the USSR created a geo-strategic vacuum, after which agents from Western countries began their encroachments as America and Russia competed for the region's resources. The new geopolitical and strategic situation invited schemes that maximised anticipated benefits by influencing decision-making processes in the region's evolving legal regime. This chapter examines the history of this legal regime with a focus on Iran and Russia. Reasons for the failure of the new littoral countries to achieve a comprehensive agreement are also explored. Moreover, we review the ongoing 'Russia versus America' rivalry that has contributed to the failure to attain a comprehensive Caspian Sea legal regime.

4.2 Sea or Lake?

There is disagreement over whether or not the Caspian Sea should be regulated as a high sea or as a lake. Some have called it a 'sea' and UNESCO's Oceanographic Commission endorsed this concept from an oceanographic perspective on the region's combination of water, animal and plant life. However, others express an opposing position, believing it to be a lake. Although contention continues, a distinct conclusion that determines the status of this broad stretch of territory is still pending. Nevertheless, it was indeed called a 'Sea' before the 1982 Convention on Sea titles based on a bias that leans towards maximizing benefits.

However, and in fact, the Caspian Sea meets the category of 'closed waters' in terms of international law, which therefore scientifically qualifies it as a lake. It is connected to the Black Sea only by man-made channels via the Volga River and lacks any natural passage to the high seas. Although historically called a 'sea' based on its vast expanse, most geologists know the Caspian Sea as a lake. It was considered 'closed waters' by the 1982 United Nations Convention on the Law of the Sea because, in accordance with this convention, a closed or semi-closed sea means a gulf, which is a basin or body of water (surrounded by two or more states) connected to another sea only through a narrow, low-width waterway (Heidari-Manesh 2009: 71).

The legal regime of seas and international waterways is a collection of principles and rules that govern relations between those who follow international maritime laws regarding internal waters, territorial seas, the monitoring area, exclusive-economic zones, high seas, the exercise of sovereignty over the seas, as well as other international waterways such as rivers, straits and international channels. Maritime Laws also pertain to ownership as well as boundaries of littoral countries over surface, bed, and claims to subsoil resources. However, the legal regime of a 'closed sea' is not subject to International Maritime Law. Therefore, its littoral countries share the sea between them and determine respective portions of ownership, as well as their own territories and approaches to the exploitation of natural resources contained on its surface and in its bed and subsoil, based on mutual agreements.

Whether the Caspian is a sea or lake is an issue related to geological science. Terms and features of concern include size, amount of salt, minerals, depth, duration of existence and formative process and/or the presence of a plateau. According to these criteria, Azov, a 38,000 km^2 area that is 14 m deep, ranks as a sea due to specific features. However, the Caspian Sea, despite its 400,000 km^2 area and 25-m depth, is categorised as a lake as formally reflected in numerous public and international legal documents (Ahmadi-Pour 2001: 42).

As for sea level, it must be said that a sea's water level is as deep as an ocean, while water levels for lakes can be deeper or shallower than an ocean's or have surface levels that are higher or lower than 'sea level'. The reason for this is that lakes are encompassed on all sides and influenced by distinctive environmental factors. The Caspian Sea level is 24 m below sea level. Hence, if the Caspian were a true sea, there would be no difference. Therefore, using 'sea' to describe the Caspian is solely related to its size. But because it has no natural passage to high seas, the Caspian falls into the category of a lake.

4.3 Historical Background of the Caspian Sea's Legal Regime

When studying the legal background of the Caspian Sea over the last 400 years, we concluded that Iran did not successfully exploit the Caspian Sea. However, Russia, due to its technological advantages, has actually used this sea for an extensive period of time (Kamran et al. 2010: 27–28).

In 1723–25, after the turmoil caused by the Afghan insurgency, the Russian Empire annexed northern Iranian borders along with that portion of the Caspian Sea. In 1723, the Russian army disembarked from a number of ships on the sea's southern coasts at the port of Anzali and seized Guilan. At the same time, a separate force occupied Baku. Following concurrent Russian ravages and Ottoman attacks, the Shah of Iran had little choice but to ask for help while surrendering to one of the invading parties. He chose the Russians and sent an envoy to negotiate peace. The agreement, signed by a representative of Tahmasp II, allowed the Russian Empire to annex occupied states as well as the Mazandaran and Astarabad states (Riyahi 1990: 59). In this way, the Caspian Sea became totally surrounded by Russian dominions. Russia even petitioned an Ottoman concord that ensured the new order. Nevertheless, Russia failed to remain in occupied coastal territories after the return of Iranian power during the reign of Nader Shah Afshar. He employed the British agent, Dalton, to undertake shipbuilding with the intent to establish naval security in the Caspian Sea. To accommodate this, he signed the Rasht contract in January 1732, and another in Ganja on 10 March 1735. Under these terms, Russia retreated from the occupied territories, including the southern coast along the Caspian Sea, 'little by little' (Damirchilu 2007: 163–164). Interestingly, not a hint on how to apply any legal rule over the Caspian Sea was in evidence, nor was there any mention of the sea's division in these eighteenth-century treaties—i.e., contracts signed in the Safavid period (1723 and 1726), and those signed during the reign of Nader Shah Afshar (1732 and 1735).

The Qajars then gained power, and when Agha Mohammad Khan Qajar retook Tbilisi in 1795, a new period of conflict between Iran and the Russian Empire began. In 1798, when Agha Mohammad Khan was killed, the renewed strife came to a head. After the succession of Fath-Ali Shah, Iran remained chaotic due to the former Shah's incessant military campaigns. Russia then re-annexed Tbilisi and attacked other cities. These wars lasted ten years and ended with the imposition of the Gulistan Treaty on Iran, whereby Iran recognised Russian sovereignty over occupied provinces. Hence, Dagestan and Georgia, along with the cities of Baku, Darband, Shirvan, Karabakh, Sheki, Ganja, Muqan and upper Talesh were ceded to Russia. In addition, Iran was divested of navigational rights in the Caspian Sea.

Considering these developments, the first treaty between Iran and Russia that determined the legal state of the Caspian Sea was the Gulistan Treaty, signed after the first ten-year war between Iran and Russia upon Iran's defeat, 12 October 1813 (Asadi-kiya 1995).[1] Before signing this treaty, no other legal documents included

[1]It should be noted that the eighteenth-century contracts cannot be considered official contracts because they did not have the features of an official document. See (Asadi-kiya 1995).

points regarding use of the sea (Kamran et al. 2010: 27–28). Hence, the Gulistan Treaty can be regarded as the historical beginning of a compilation of contractual laws over use of the Caspian Sea. Following yet another war, the Turkmenchay Treaty confirmed this beginning in 1828. The Gulistan Treaty was unclear regarding borderlines between Iran and Russia, which also gave cause for Russia to make later forays into Persian territory, including lands around Iravan and Azerbaijan. The oversight allowed a second round of conflicts from 1826 to 1828, finishing with the Turkmenchay Treaty. Accordingly, Iravan, Nakhjavan and part of the Muqan plain were ceded to Russia in addition to lands previously transferred by the Gulistan Treaty. Iran was further obliged to compensate Russia with five million Toman. The capitulation of maritime rights and consular jurisdiction was also imposed on Iran. Hence, under the Gulistan (1813) and Turkmenchay (1828) treaties, not only was the western coast along the Caspian Sea lost to Iran, but limitations were also placed on Iranian rights in the Caspian Sea. Therefore, and for the first time, the rights of two countries were explicitly mentioned in documents regarding the Caspian Sea. Article 4 of the Gulistan Treaty, in addition to referring to the parties' free access to each other's ports and the freedom of merchant vessels for both parties, clearly stipulates that only Russia reserved the right to a military presence in the sea. The same article was reiterated in the Turkmenchay Treaty, signed after the second Russian-Iranian War (Ögütçü 2003: 36).

According to Chapter V of the Gulistan Treaty, and to Chapter VIII of the Turkmenchay Treaty, the rights of both countries were cited with regard to shipping in the Caspian, along with a prohibition of any Iranian military presence (navy) in the sea. Chapter V of the Gulistan Treaty reads as follows:

> Russian government ships trading on the Caspian Sea, due to the former command will be permitted to depart from or approach the coast and ports of Iran and in the case of storm or shipwreck, Iran shall render them friendly assistance. On behalf of the former command, the ships of Iran shall also be permitted to approach Russian coasts for transactions and similarly, in the case of storm or shipwreck, they shall be rendered friendly assistance from the Russian side. Russian public war ships, in peace or any other times when under the flag of Russia, have also been given brotherly permission to travel as mentioned in the former command, and no other state shall have warships in the Caspian Sea but Russia.

Chapter VIII of the Turkmenchay Treaty states:

> ARTICLE VIII: Russian trade ships still preserve the right to freely sail in the Caspian Sea and along its shores, as well as to approach them. In the case of a shipwreck Iran shall render assistance to them. The trade ships of Iran also have the right to sail in the Caspian Sea and to approach the shores of Russia and in case of a shipwreck Russia shall render any and all necessary assistance to Iran. As to public vessels, as stated before, they have the right to sail only under Russia's flag. They preserve the former exceptional right and no other country may have public vessels in the Caspian but Russia.

In this regard, the nineteenth century can be considered the beginning of a new chapter in the history of the Caspian Sea and its basin, as well as the beginning of the decline of Iranian shipping and the beginning of its naval exclusion from the Caspian Sea.

Following the social and political events of Russia's October Revolution (1917), dramatic developments changed globally and regionally. After the Tsar's demise, the legal status of the Caspian Sea, as drawn by the Gulistan and Turkmenchay treaties, changed. Russia proceeded to cancel all past colonial contracts with its neighbours while consolidating its revolution. Only in 1921 did Russia enter into new agreements with neighbouring countries. The Amity Treaty of 1921, between Iran and Russia, then became a new basis for the Caspian Sea's legal regime. Under this treaty, all previous agreements were annulled (Mirfakhrai 2005: 184). Its Article 11 emphasised equal rights with the Soviet Union for shipping and the conclusion of contracts, both military and commercial. Articles 13 and 14 pertained to fisheries where Russia asked Iran to grant fishery rating to Russia for Russian livelihood. Iran agreed (1925) and signed a rating contract for a period of twenty-five years (Bavand 2008: 9).

After the Treaty of Amity (1921), other agreements between both countries followed. The Sailing Contracts of 1940 (25 March 1940), signed in Tehran, are the most important as they clarify general positions for both states regarding shipping and fishing, but remain silent regarding other issues, mainly because the Sea's unseen resources (oil and gas) were not yet discovered. Chapters 7 and 11 of the 1921 Act, and Articles 12 and 13 of the 1940 Convention regarding navigation and commercial ship monitoring in the Caspian Sea are based on the principle of equality. Russia imposed unilateral sanctions on an Iranian naval presence in the Caspian Sea via the Amity Treaty (1921), and prevented Iran from employing foreign nationals with hostile intentions towards Russia (Navazani 1991: 1). This limitation was a result of Russia's own refusal to allow the presence of foreign nationals in the Soviet Fleet rather than any Soviet tendency to apply military superiority. These deductions are based on the content of letters from Foreign Minister Muzaffar Azam's letter (25 March 1940) in response to the USSR ambassador in Iran.

Articles 12, 13 and 14 of the Commerce and Sailing Treaty (26 February 1921) between Iran and the USSR are very important. For the first time, on the basis of Clause 4, Article 12, a distance of ten miles from the coastline of each country was recognised as an exclusive fishing zone. Apart from this provision, each party pledged to assign fishing rights in respective coastal waters up to ten nautical miles, and each country reserved this right for the benefit of fish imports caught by commercial fishermen, under respective flags, for discounts and other special benefits. Furthermore, according to Article 13, both parties, in accordance with the principles of the Treaty, agreed that, except for Iran and the USSR, ships owned by citizens and commercial businesses for transportation from either contracting party under flags of Iran or the Soviet Union, other national flag carriers could not traverse the Caspian Sea. Later, the 1940 Treaty clearly stipulated that the Caspian Sea is a communal sea shared by its littoral states, by which each party has equal rights to use the sea as permitted by themselves. The Sailing contract apparently set the principle of equality as an important legal precedent in the field of shipping (Kamran et al. 2010: 28).

Despite these agreements, Iran's Caspian Sea shipping and fishing activities remained extremely limited, due to a lack of technical facilities and political will, considered major causes (Damirchilu 2007: 165). What is reasonably presumable by these cited articles is that Moscow and Tehran agreed that Caspian Sea routes should be closed to the rest of the world and remain exclusively their own. Aside from the 'ten nautical mile' exclusive fishing right, there were no other attempts made to determine any other maritime boundaries, nor did these treaties mention anything about the development of seabed mineral resources. Thus, the Caspian Sea was administered as an enclosed 'inland water' asset, shared communally between Iran and the Soviet on a joint-formal basis, as per the 1921 and 1940 treaties, prior to the collapse of the Soviet Union. For this reason, many analysts and policy-makers question the applicability of these documents to the post-Soviet era (Bahgat 2005: 265–271).

After the Soviet collapse, a group of Iranian lawyers appealed for fifty per cent rights to Caspian Sea resources. They claimed this fifty per cent Iranian right was directly related to legal precedents between Iran and Russia that granted equal rights to shipping in the Caspian Sea. Some contemporary pundits claim equal rights to any division of the Caspian Sea's seabed and subsoil. Nevertheless, although the USSR recognised the Caspian Sea as a common lake shared with Iran on the basis of the cited treaties, the Soviets never respected Iranian rights in sectors related to shipping, land or maritime territory in the Caspian Basin. For example, Iran was denied any military presence in the Caspian Sea and could not devote any such resources to oil exploration or extraction on its shores in Mazandaran (Mojtahed-Zadeh 2009: 39). Generally, this so-called 'legal status' in the Caspian Basin during the Soviet era can be considered 'equality in opinion and inequality in practice'. However, even this precedent cannot be taken into account when establishing any acquired rights by the Soviet Union (Damirchilu 2007: 166).

During the time that Iran and the Soviet Union were the only littoral states of Caspian, their bipolar mini-world with its political and military priorities and extremely secure borders bore many economic and non-economic benefits. In the shadow of several decades of tolerance and forbearance, the region enjoyed stability and cooperation and neither party felt a need to further any determination of respective legal rights. But this drastically changed with the addition of three new littoral countries: Azerbaijan, Turkmenistan and Kazakhstan. The two traditional partners and their mutual provision and entry into independent activities in the basin had to be reset, especially regarding fishing, shipping and the race to exploit new-found oil and gas resources. Hence, fundamental changes in attitudes, geopolitics and international circumstances demanded that the legal status of the Caspian Sea be attended to as a priority for all beneficiaries (Bavand 2008; Kumar 2009: 13).[2]

After the collapse of the Soviet Union, coastal states began several levels of negotiations to resolve Caspian Sea disputes. Iran took the first step in 1992 at

[2]After the collapse of the Soviet Union, according to International conventions on government succession, newly independent littoral states must accept past Iranian obligations, especially in relation to territorial (border) contracts and communal waters. Therefore, Turkmenistan, Kazakhstan and Azerbaijan announced at the Almaty Conference that they would adhere to the obligations of the former Soviet Union but later adopted a different policy. See Bavand (2008: 10). Consequently, and considering the extent of the Caspian Sea's coastline, they are attempting to redefine the legal status of the Caspian Sea as a sea rather than a lake. See Kumar (2009: 13).

a summit for member states of the Structural Economic Cooperation (ECO) and proposed the establishment of a Caspian Sea Cooperation Organization by inviting Russia's representative to the meeting. Nevertheless, unilateral actions taken by the newly established littoral nations on the use of hydrocarbon resources turned towards the incorporation of trans-regional powers such as the US and Western oil companies. Naturally, this led to divergence rather than regional convergence. Subsequently, five-sided negotiations leading to cooperation and bilateral agreements between Russia, Kazakhstan and Azerbaijan have tentatively divided the Caspian Seabed for the exploitation of its mineral resources as follows:

1. The 'Kazakhstan and Azerbaijan Agreement', based on common commitment which defined boundaries on the basis of a so-called 'Fair line' (1997);
2. Agreements between the Russian Federation and Kazakhstan on the division of the northern Caspian Sea basin that apply rules for the use of underground resources (16 July 1998) and the implementation of a later set of supplementary protocols dated 13 May 2002;
3. Agreements between the Russian Federation and Azerbaijan on the division of adjacent areas of the Caspian Seabed, dated 23 September 2002;
4. Agreements between Azerbaijan and Kazakhstan on the division of adjacent regions of the Caspian Seabed, dated 29 November 2001, and the implementation of protocols, dated 27 February 2003;
5. Agreements between the Russian Federation, Azerbaijan and Kazakhstan on the junctions of adjacent areas as dividing lines for the Caspian Seabed, dated 14 May 2003 (Damirchilu 2007: 195).

Despite these agreements, all five governments have remained in conflict over Caspian maritime borders since 2002. Thus far, all Caspian Sea negotiations have yielded little result and these littoral states have not avoided the Caspian Sea's militarization as a consequence. Furthermore, unilateral actions taken by individual countries to exploit mineral resources, especially oil and gas, have exacerbated tensions. In the Baku meeting, contemporary parties apparently agreed on a scheme on how to approach a viable legal convention for the sea. However, a week after the meeting, the Iranian government rejected this experiment by stating they did not recognise the validity of Russia's bilateral agreements with Azerbaijan and Kazakhstan on the sea's delimitation (Blank 2011). Currently, only Russia and Kazakhstan have established common borders sufficient to exploit the sea's reserves. The three remaining countries still hold mutual disagreements. For example, Tehran and Baku disagree on the reserves at the Plum, Shargh and Araz oil fields. Hence, despite any bilateral conclusion to agreements that furthered respective interests, Caspian littoral states adopted independent positions and *modus operandi*.

The most important issues discussed in their negotiations are given below:

1. A conventional determination of the legal regime, including limits related to sea levels, governance, economics, fisheries, communal waters, and the delimitation of the seabed and subsoil;
2. Security and law enforcement and bans on non-littoral participation;

3. Energy resource exploration, exploitation and construction of pipelines;
4. Environmental concerns;
5. Transport, shipping, ports and free transit through the Volga-Don Canal;
6. Trade and Economic Cooperation.

The more important issues are legal regime conventions, delimitation, military, and security and law enforcement issues. These discussions, along with disputes, continue and some differences remain unresolved. Consequently, the five littoral states have failed to design a common framework for sustainable development of the basin. However, since the sea's geographical enclosure dictates that each country cannot specifically define respective national regions, and despite numerous definitions and interpretations, they all agree that the legal regime should be developed by consensus. Hence the establishment of the Caspian Sea Cooperation Organization and the five committees that assist the basin's administration:

1. The Committee of Scientific Research to monitor geological fluctuations;
2. The Committee for Caspian Sea Environment Protection;
3. The Committee of Biological Resources;
4. The Committee for the Legal Regime;
5. The Committee of Transport (shipping and resource use).

4.4 Proposals to Divide the Caspian Sea

Based on discussions that occurred after the Soviet collapse, several proposals to divide the Caspian Sea have been put forth but without agreement. These include the following:

- *Joint or Condominium Method*:
  Initially proposed by Russia and later considered by Iran and Turkmenistan. Until 1998, the Russian government supported a regime based on 'common exploitation' as its official stance. Each of the five countries would be involved in the reserves of the Caspian Sea's continental plateau as a communal regime according to international standards. Parts of the continental plateau closer to respective shores would be especially reserved for each country. What remained would then belong to all five countries.
- *Connection Method*:
  This proposes that the final contact point of land borders would be connected. Hence, all waters behind this point become a monopoly of that country. Application would severely lessen western and eastern Caspian countries from participation in the lake. At the Almaty session, Russia emphasised the delusion of this approach.
- *Division by focus*:
  This calls for determining the epicentre of the basin followed by land borders connecting with this determined centre, which would mostly benefit Iran and Kazakhstan.

- *Division* via *two foci*:
  Supposing the sea was oval, two foci are determined and connected. Then, by extending land borderlines to these foci, the sea becomes divided into five parts. In this case, Iran's share would be twenty per cent. The proposal was rejected.
- *By using the most advanced land areas into the sea:*
  To implement this proposal in the southern sea, a straight line connects three points, the Sefidrud Estuary, the Copa peninsular headland, and the southern headland of Krasnosk Bay in Turkmenistan. Perpendiculars from these lines are then drawn. The intersection of these perpendiculars is determined and lines are drawn to the border endpoints of the three countries. This method favours Kazakhstan whose portion would be greater than Russia's, with remaining shares divided equally between Turkmenistan and Azerbaijan.
- *Division on the basis of coastline length*:
  Here, a fair north-south line is drawn through the sea, and, proportionally to shore length, extensions are drawn to land borders from the fair line.
- *A Communal Regime*:
  Division into five equal parts with equal share distribution for each state for the sea, seabed, and sub-seabed: this method draws land border lines into the sea in such a way that five equal parts result.
- Increase of national ownership boundaries to fifteen nautical miles and the merging of two coastal areas of the border for customs' control with an exclusive fishing area—Russia's suggestion at the Tenth Summit (July 2003, Moscow).
- Commitment to both 1921 and 1940 agreements:
  Both agreements form a foundational framework for the lake's legal regime for both present and future—Iran's suggestion at the Eighth Summit of the Caspian Working Group (March 2003, Baku).

These suggestions are littoral state positions and present competition as a major feature, with each seeking to gain a greater share (Etaat et al. 2012: 129).

4.5 A History of Littoral Perspectives on Drafting a Legal Regime

Russia: Immediately after the Soviet collapse, Russia was accorded condominium status and moved in protest against Azerbaijan's proceedings regarding exploitation and division of the lake. The 'midline' adjustment changed the previous procedure and was based on a 1998 formula. Accordingly, bed and over-bed extending ten miles from the coastline was considered a 'national coastal zone'. Portions of the remaining coastal seabed for each littoral country were divided according to the 1982 Sea Convention's legal regime governing lakes. This division implied that the midline was defined as the seabed's boundary for each of the littoral countries. The midline is also adjusted for sea bulges, including concave and convex states, and artificial coastal facilities. The formula recognised the seabed, beyond the national ten-mile zone, as a communal region (Omidi 2006: 183). These rules meant that any underground resource in divided sections belonged to the respective country, but seabed beyond sovereign domains did not, and furthermore, it allowed for unrestricted submarine passage in communal zones.

Although Russia initially subscribed to the condominium position, after the Soviet collapse and Caspian negotiation failures they contracted with Kazakhstan and Azerbaijan to define boundaries. Experts believe the Russian position changed due to the discovery of oil and gas resources. Previously, Russia defended the communal perspective when no oil or gas resources had been discovered on the banks of either country. Currently, Russia advances maximum exploitation facilitation because of its coastal proximity and high degree of technology regardless of any legal regime/division. Russia's LUKOIL Company widely invested in the Cochin area in April 2010, as the most important oil project in the Caspian Sea.

Kazakhstan: Kazakhstan claims that because the Caspian Sea has links to the Baltic and Black Seas through both the Volga and Don Rivers, it has the property of a high sea. Therefore, it believes the legal regime derived by the 1982 Convention on the Law of the Sea should apply. This means drawing baselines of twelve nautical miles as territorial waters, twenty-four miles in adjacent areas, and 200 miles as exclusive economic zones on the continental plateau. Kazakhstan insists that the Caspian is a 'High Sea', conforming to international laws governing freedom of navigation, flight, as well as the deployment and placement of pipelines and cables. Regarding the area of the sea, the Kazakhstan perspective is exclusive. It claims restricted rights to seabed resources in its own zones, as just defined, with the rest of the water used jointly with Russia. Thus, it would share access rights to the Volga and Don canals.

As noted previously, Kazakhstan, Russia and Azerbaijan had divided the seabed between them. However, Kazakhstan believes that a more accurate determination between littoral countries should be based on bilateral or multilateral agreements and suggested that the division should be based on a median line according to bisects for adjacent countries. Kazakhstan also successfully achieved some practical implementation. For instance, in 1998 it signed a bilateral agreement with Russia. Accordingly, the northern sea was divided between Kazakhstan and Russia with defined boundaries. Hence, the Kurman Ghazi field, owned by Kazakhstan, was recognised by Russia (Bijani 2003: 159).

Regarding the 1921 and 1940 contracts, Kazakhstan believes that agreements preceding the Soviet collapse are invalid and no longer stand as legal foundations. In fact, Azerbaijan and Kazakhstan are both believed to have divided the Caspian Sea unilaterally. However, Kazakhstan's perspective is unique compared to Azerbaijan's position, in that it wants the sea divided according to the 1982 Convention on the Law of the Sea. Conversely, Azerbaijan, as a former Soviet republic, wants the sea divided according to former Soviet boundaries. It must be stated that thirty to forty per cent of Kazakhstan's oil reserves are located in the Caspian Sea, and Kazakhstan can only utilise these resources if the sea is divided between the littoral countries (Torabi 2002: 67–68).

Turkmenistan: Turkmenistan's position is mutable and ambiguous. Initially, it supported Russian and Iranian positions. At the same time, it passed a law (October 1993) stipulating a twelve-mile exclusive economic zone (Mehdiyoun 2000: 183). In 1996 however, Turkmenistan and Iran proposed a forty-five-mile area zone for the use of abiotic resources, with communal usage designated for the remaining seabed (Ghafouri 2008: 88). Then, in 1997, it supported a seabed division theory based on a median line, declaring a normative agreement alongside Kazakhstan (Mehdiyoun 2000: 183). Turkmenistan also supported borders presented in the *Soviet Union Times* (Ghafouri 2008: 88). Meanwhile, on 8 July 1998 it supported Iran's idea of a communal sea and announced '*the 1940 and 1921 treaties are in force since the final position of the Caspian Sea legal regime is determined*', and since that time the sea should be used in a communal way. It further said that if it is divided, it should be divided into five twenty per cent shares and that the seabed and water required a unique regime.

Turkmenistan's position on the legal regime of the Caspian Sea changed during the years after the Soviet collapse. At first it defended the seabed's plan for division. In a joint statement with Kazakhstan (27 February 1997), both countries advocated a midline-based division of the seabed and sea until a perfected legal regime was established. Turkmenistan believed in the past pact between Iran and the Soviet Union, and also the sharing of the USSR with former Soviet Union republics within a domestic judicial framework. Hence, Turkmenistan, much like the Republic of Azerbaijan, approved a law in this sense. Although Turkmenistan's national borders naturally opposed any division of the Caspian Sea, when it objected, along with Iran, to bilateral agreements, it became clear that Turkmenistan also wished to divide the Caspian Sea because its proposals obviously implied inevitable division. Hence, it seems likely that if its conflict with Azerbaijan over the ownership of Kapaz and Osman fields is resolved, Turkmenistan will also consent to a complete division of the Caspian Basin. In fact, it can be surmised that it disagrees with littoral countries on 'how to divide' rather than the principle of division (Torabi 2002: 69).

The resources within its continental plateau include thirty to forty per cent of its total oil and gas reserves, with most of the production along Turkmenistan's near-shore marine areas. Thus, the legal regime Turkmenistan supports makes it possible to seize the cited continental plateau resources. In 1997, it specifically signed a bilateral agreement with Azerbaijan to divide this oil field. However, subsequent studies showed that the bulk of this field's reserves were located within Azerbaijan. Therefore, Turkmenistan abolished the agreement. From then on, disagreements between both countries over oilfields have intensified. Thus, Turkmenistan's support of the 1921 and 1940 treaties did not mean that the country's politicians believed in a communal legal regime, but rather aimed to avoid Azerbaijan's exploitation of the disputed oil fields.

Azerbaijan: Azerbaijan can be considered the prime Caspian Sea legal regime disrupter. Azerbaijan was the first country that ignored the legal regime of the Caspian Sea by signing the initial oil contracts with several Western oil companies in September 1994. Primarily concerned with its national interests, Azerbaijan wished to acquire the greater portion of the Caspian Sea's oil and gas resources. Hence, it agrees to a complete division based on the equitable line. Azerbaijan claims that the Caspian Sea is an international sea or border lake with a seabed and adjacent coastal waters on each state and should be divided so that each state has full control over its own section (Mamedov 2001: 221). This division should be based on principles of international division according to a median line. With regard to emerging countries, its concept of the division is based on the dividing line drawn by the Soviet Union Oil Ministry in 1970, which was specifically designed for oil exploration and extraction operations for the former USSR Republics. Under such a division, shares for Azerbaijan, Kazakhstan, Turkmenistan, and Russia would be 70,000, 105,000, 80,000 and 60,000 km^2, with the remaining 55,000 km^2 for Iran (Ghafouri 2008: 87). The Azerbaijan perspective, assuming that Russian-Iranian treaties are valid, assumes they are not authorised to extract oil and gas. Regarding these treaties, Azerbaijan claims the treaties are not acceptable due to changed circumstances after the Soviet collapse because free independent government policy and the cited legal regime are mutually exclusive. It also claims that the use of continental plateau oil resources since 1956 has not been contested by others. However, some experts believe that Azerbaijan's argument is incorrect because Azerbaijan initially accepted the validity of former treaties between Iran and the Soviet Union as per the Almata Declaration, and its proposed model for Caspian Sea division implies unequal benefits for littoral countries of a communal sea. Critics also say Azerbaijan's lack of attention to continental plateau oil resources at the signing of the 1921 and 1940 treaties was because exploitation of these resources was neither discussed nor mentioned in any bilateral or regional international agreements before 1945. Therefore, critics posit that the applicable legal regime governing the 1921 and 1940 treaties concerned all aspects—citing also that the administrative internal borders of the former Soviet Union are not binding on third countries, including Iran (Shariat Bagheri and Jahani 2001: 147).

Iran: Iran's interests in the Caspian Sea had been violated by the former Soviet Union for decades. The disintegration of the Soviet Union provided an opportunity to claim Iranian rights and promote its historically valid national interests in the Caspian Sea (Mojtahed-Zadeh 2009: 83). Contracts concluded between Iran and Russia (1921 and 1940) specified the sea's management before the Soviet collapse. Hence, Iran desired a communal approach, because those contracts referred to the Caspian Basin as an exclusively shared Iranian and Russian territory. Principles arising from these contracts comprise two themes required by a condominium regime; i.e., equality and exclusivity. In agreement with Russia's initial position, Iran insists the Caspian Sea is a communal water basin. Hence, Iran initially preferred that a portion of the sea be reserved for respective national monopolies with the balance, including the seabed and sea, considered a communal area. But after the shift in the Russian position, Iran wished to divide the sea equally between the five littoral countries, giving each a twenty- per cent portion. The latter position faced opposition from Russia, Azerbaijan and Kazakhstan, who considered Iran's plan an impractical suggestion and then countered with arguments positing proportional distributions based on respective coastal sizes. However, the Iranian position satisfied Turkmenistan. Nevertheless, after acknowledging its impracticality, Iran changed positions and called for a complete division of the sea. Believing that other criteria qualified as equity apart from coastal size, and taking into account Iran's coastal shape, the applied mathematical formulas projected Iran's share at twenty per cent (Omidi 2006: 184–185). Even so, the littoral countries also rejected this because they believed it totally unjust, since there was little to no correlation between respective shares and coastal size. The Russians then introduced a new proposal to escape the deadlock. Here the sea was divided into 'forty-nine per cent north' and 'fifty-one per cent south' by calculating a midline. The southern portion was divided into equal parts between Iran, Turkmenistan and Azerbaijan, each share equal to seventeen per cent This proposal was contested by Azerbaijan and Iran.

In general, Iran insists that the legal regime of the Caspian Sea should be achieved through the consensus of all littoral states and that the treaties of 1921 and 1940 should remain valid until an agreement on a perfected legal regime is reached. Iran further believes that any unilateral use of resources, especially those of the seabed, is unacceptable. Iran has also denied the validity of bilateral agreements and objected to those between Kazakhstan and Russia, and Russia and Azerbaijan. In fact, Iran has defended its twenty per cent share from the beginning of discussions on the sea's division, although such a classification is contrary to the views of Azerbaijan and Turkmenistan (Aghai Diba 2004: 1–2).

Any division based on the coastal length faced Iranian opposition because Iran has the smallest coastline, which is concave in shape and provides a smaller portion of the blue zone than convex formulations when determining boundaries. Being a sea or lake is yet another geographical contention, which adds to Iran's support of communal use of the Caspian Sea's oil and gas resources. Therefore, Iran benefits most from oil and gas resources by insisting on the condominium perspective that grants it a minimum of twenty per cent.

In addition to low volumes of resources in coastal areas, the sea's depth makes it difficult to extract these resources.

4.6 American and Russian Roles in Determining the Legal Regime of the Caspian Sea

For purposes of hegemony, America and the Russian Federation have tried to increase their economic power and influence in the earth's key strategic regions. With the prospect of Eurasian and Caucasus hydrocarbon potentials, the Caspian Sea became a particular cause of conflict for regional and trans-regional powers. To strengthen economic power and strategic positioning, both nations have attempted to influence the Caspian Sea's legal regime. At times this occurs by economic assistance and financial strategy, and occasionally by political and military intervention (Azhdari and Qaeini 2012: 28). Economically, the Caspian Basin presents attractive opportunities for the West. Eurasia has two major energy-source foci. One is the Persian Gulf and the other is Siberia and the Eastern Urals. The Caspian Sea lies between them. Hence, it is now a necessary option for consumer-orientated countries concerned with growing conflicts in the Middle East that threaten energy production and transport.

4.6.1 The United States and the Legal Regime of the Caspian Sea

According to both the US Ministry of Energy and the International Monetary Fund, the volume of Caspian Sea oil reserves is 178 billion barrels, the greater portion of which lies off Kazakhstan's coasts, estimated at 95 billion barrels. Turkmenistan and Azerbaijan reserves are estimated at 33 and 31 billion barrels, respectively. Their reports add that the greatest gas reserves are in Turkmenistan, estimated at 8.9 trillion barrels, with Kazakhstan's and Turkmenistan's gas reserves estimated at 4 and 1.3 billion barrels, respectively. Energy analysts of the Baker Institute believe that the American government pursues other benefits besides energy in the region (Azhdari and Qaeini 2012: 30). While the littoral states try to preserve the region's independence and stability, with Russia as an equal but not a dominant partner, the US continually confronts Russian ambitions as well as religious militancy in the region by establishing free-market economics, ensuring respect for the rule of law, and creating a civil society. The American Congress adopted the 'Silk Road' strategy in March 1999 to expand its role in the Caspian region. This policy determined goals such as friendly regional relations and economic cooperation to achieve market shares; the protection of American companies in the region; and developing relations between local republics and stable democratic Western nations. America encourages Caspian countries to comprehensively utilise their resources, especially energy assets, to better enable economic upgrades for the entire region that facilitate a growing consumer market for regional American companies. America's economic prosperity as the world's largest economy is related to its increasing consumer growth and the economic convergence of important commercial poles; thus, America seeks to develop energy security while maintaining a continuum of ongoing commercial convergence (Hersani 2006: 41).

In general, energy security plays a large role in the codification of American foreign policy, to include every country with energy reserves, which then determines a country's value as a priority foreign policy concern. Accordingly, US foreign policy experts design national energy policies that provide for discontinuous energy supplies, reliable sources, and accessibility to public and economic sectors based on calculated expectations of needs, resource and reserves. After the 2001 WTC attacks in New York, American and Western nations became more attracted to Central Asian and Caucasus regions for purposes of energy security. As circumstances in the Middle East changed after exacerbations in Iraq, the West predicted that its regional and global ambitions in Middle Eastern oil-producing countries would be stricken by tension, conflict and unforeseen events. Hence, it became necessary to acquire alternative energy sources that could respond to a crisis without any interruption in energy supply. American authorities then evaluated extant resources in the Caspian Sea, Central Asia and the Caucasus as supplemental but important resources in times of changing circumstances. Such an immense scheme and long-term strategy reduced reliance on Middle East sources and can be considered in numerous ways (Fayazi 2008: 33).

In addition to the economic benefits derived from hydrocarbons, US foreign policy framework in the Caspian Sea region follows several strategic parallels that include the following major points:

- American presence and activity in newly independent states with the pretext of promoting democracy, but with the practical goal of preventing penetration, dominance, and cooperation with China, Russia and Iran;
- Easy access to oil and gas resources without potential risks, allowing the West to diversify energy supply and transport;
- Implementation of a strategy based on energy security that reduces dependence on Persian Gulf oil;
- Use of Caspian Sea energy resources and the construction of various pipelines for transport (Hersani 2006: 36–39).

US strategy also includes military options that opt for seizure of the Caspian Basin. Since 1996, the region has been considered of '*vital interest*', which places it in the Military domain of responsibility. In 1998, America attempted to resolve disputes between Turkmenistan, Azerbaijan and Kazakhstan and bring them into the 'East-West Corridor'. This effort included prompting American oil companies to invest in Caspian and Central Asian reserves, and the provision of financial grants as well as armament and military training. Such interventions delayed any resolutions towards a regional legal regime because the benefits offered were to the exclusion of major littoral contributions. The impositions prevented Russian revitalization (Ahmadian and Gholami 2010: 13), and pressured Iran towards a forbidding programme that limited its involvement as much as possible in several fields of endeavour (Azhdari and Qaeini 2012: 33). In addition, American exploitation of Caspian resources undermined ongoing regional negotiations regarding the sea's legal regime. Since energy security shares a major portion of determinants that comprise American foreign policy as discussed above, they facilitated delays in the determination of a Caspian Sea legal regime. These strategies included stimulating disagreements leading to division, and an American military presence under the pretext of fighting terrorism and Islamic fundamentalists.

Generally, America's national interests in the Caspian Basin have five goals:

- Support territorial integrity and the political independence of newly independent republics to reduce Iranian and Russian influences;
- Prevent state-failures that might otherwise encourage the hegemony of Islamic groups who oppose America's cultural impositions;
- Improve energy security by diversifying sourcing and transport while supporting democracy and 'free market' economics;

- Play a major role in the region's economic growth vis-à-vis resource exploitation;
- Establish a military presence and increase military cooperation with the region's littoral republics via programmes such as '*Partnership for Peace*' while aiding the eastern NATO encroachment (Ebrahimi and Mohammadi 2011: 6).

4.6.2 Russia and the Caspian Sea's Legal Regime

Regional security has been the most important aspect of Russian foreign policy. After the collapse of the Soviet Union, Russia's southern borders were viewed as a security matter and any foreign influence was/is considered a threat. In view of traditional-geopolitical concepts, Russia deems its southern borders an exclusive sphere of economic, political, and military influence. Hence, any entry of foreign players was regarded as an imminent threat to Russian national interests (Godzimirski 2008: 21).

Russia holds that its rightful interests in the Caspian Sea are legitimate for three main reasons. First, Russia is one of the five littoral countries; second, despite the USSR's collapse, the Russian economy is intimately associated with the new republics. Hence, any changes can have positive or negative impacts on its economy. Third, Central Asia and the Caucasus, for both Eastern and Western powers, are geopolitically considered a Russian territorial security matter. Therefore, any foreign presence directly impacts Russian security (Shouri 2007: 42).

Russia continues its own approach to imperialist hegemony in direct competition with the United States; in doing so, it also stands in urgent need of energy security to strengthen and maintain its position. Hence, Russia is specifically focused on Caspian Sea resources with exploitation in mind. In line with its strategy, especially in the case of hydrocarbon resources, two major policy themes arise:

- Resource extraction: The Caspian Sea's major oil and gas reserves are located within the geographical boundaries of both Azerbaijan and Kazakhstan (oil), and Turkmenistan (natural gas). All three countries lack essential technical and financial means for resource exploitation.
- Energy-transport must be considered to transfer this energy to consumers, as the Caspian Sea has no passage to the high seas. Kazakhstan, Azerbaijan, and Turkmenistan are otherwise landlocked countries.

Russia, America, and other powers are looking for ways to reap the greatest benefits. Russia believes any extraction and transportation efforts made by others are an investment of foreign influence accompanied by security risks. Hence, it undertook a large military exercise in spring 2002, specifically intended to highlight Russian military power in the region and, perhaps, impose its own legal regime on its neighbours, although this is speculation.

In addition to all the above, the Caspian Sea holds a set of security priorities for Russia that include:

- Central Asia and the Caucasus are major foci of Russia's foreign policy. Russian leaders have looked at 'close overseas' and security as primary foreign policy concerns and expect other states to recognise their singular interests in these regions. Russia holds itself especially responsible for securing order in 'close overseas' and plays an actively constructive role in the development of future military and political regional relations (Koulaei 1997: 221–222).

- Russia needs the stabilization of political and military balance in the newly independent countries. Stabilization and preventing armed conflicts near Russia's border are of the utmost importance, especially out of concern for the spread of any regional turmoil to Russian soil; hence, the concern over security (Mojtahed-Zadeh 1999: 159).

Strategically, the Caspian Sea is Russia's backyard. Many people wish Russia, as the largest surviving state post-Soviet collapse, to maintain power and. under appropriate conditions, grow in strength. Accordingly, and more than any other nation, Russia rightfully participates in processes holding significant implications for the entire region, including the determination of a legal regime for the Caspian Sea. Russia shares common interests in the Caspian Sea and it does its utmost to access and maximise benefits from this sea. After repeated signings of unilateral contracts with neighbouring countries, Russia exploits Caspian's resources for its own interests, which include having as many shares as possible, especially since it is the only littoral country with the ability to do so.

References

Aghai Diba. B. (2004). The legal regime of the Caspian Sea: Any changes in the position of Iran? Online at http://www.payvand.com/news/04/feb/1169.html. Accessed Nov 25, 2017.

Ahmadian, Q., & Gholami, T. (2010). Asiya-ye Markazi va Qafghaz dar Arse-ye Taaroz-e Manafe-e Rusiye va Qarb (Central Asia and the caucasus as a conflict area of interests between Russia and the West). *Motale'at-e Orasiya-ye Markazi (Central Eurasian Studies): Center of International Studies, Faculty of Law and Political Science*, *2*(4), 13.

Ahmadi-Pour, Z. (2001). Jeopolotice Iran ba'd az Jange Sard (Iran's geopolitics after the Cold War). *Pajouheshhaye Joghrafiyai Journal, 39*, 42.

Asadi-kiya, B. (1995). A glance at the legal status of the Caspian Sea. *Evening Daily*, 16.

Azhdari, B., & Qaeini, H. (2012). Naqsh-e Amrica va Rusiye dar Tain-e Rejim-e Hoquqiye Daryay-e Khazar (America and Russia's role in determining the legal regime of the Caspian Sea). *Domahnae-ye Tahavolate Iran va Orasiya (Journal of developments in Iran and Eurasia), 35*, 28.

Bahgat, G. (2005). Oil and terrorism: Central Asia and the Caucasus. *The Journal of Social, Political, and Economic Studies, 30*(3), 265–281.

Bavand, D. H. (2008). Barresi-ye Rejim-e Hoquqi-ye Darya-ye Khazar Dar Mosahebe ba Doctor Hermidas Bavand (The legal regime of the Caspian Sea in an interview with Dr. Hermidas Bavand). *Gozaresh, 17*(192), 9–12.

Bijani, M. (2003). *Khazar va Manafe'-e Jomhouri-ye Islami-ye Iran (Caspian and Interests of the Islamic Republic of Iran)*. Tehran: Markaz-e Motale'at-e Sepah.

Blank, S. (2011). Caspian basin: Frustration mounting, Moscow talks about militarizing. *Eurasianets Weekly Digest*. At: http://www.eurasianet.org/node/63470. Accessed on Dec 12, 2017.

Damirchilu, M. (2007). Nagah-e Iran-i be Syr-e Tahavolat-e Rejim-e Hoquqi-ye Darya-ye Khazar (Iranian point of view on the evolution of the legal regime of the Caspian Sea). *Faslname-ye Motale'at-e Asiya-ye Markazi va Qafqaz (The Journal of Caucasus and Central Asian Studies), 4*(53), 159–186.

Ebrahimi, S., & Mohammadi, M. (2011). Reqabat-e Rousi-ye va Amrica dar Qafqaz-e Markazi (1991–2010) (Russia-South America Rivalry in the Southern Caucasus). *Motale'at-e Orasiya-ye Markazi (Central Eurasian Studies), 4*(8), 6.

Etaat, J., Nusrati, H. R., & Varzesh, I. (2012). Avamel-e Mo'aser dar Adam-e Ta'in-e Khotout-e Marzi-ye Daryai-ye Keshvarha-ye Saheli Darya-ye Khazar (Effective factors in the lack of determining the maritime borders of the Caspian Littoral Countries). *Faslname-ye Rahbord (Journal of Rahbord), 20*(61), 129.

Fayazi, M. (2008). Reqabat Piramoun-e Enerji-ye Asiya-ye Markazi va Qafqaz (Competition on the energy of Central Asia and Caucasus). *Faslname-ye Asiya-ye Markazi va Qafqaz (Journal of Central Asia and Caucasus), 60,* 141–168.

Ghafouri, M. (2008). The Caspian Sea: Rivalry and cooperation. *Middle East Policy, 15*(2), 88.

Godzimirski, J. M. (2008). Putin and post-soviet identity: Building blocks and buzz words. *Problems of Post-communism, 55*(5), 14–27.

Heidari-Manesh, A. H. (2009). Rejim-e Hoquqi-ye Darya-ye Khazar va Konvansion-e Hoquq-e Daryaha (The Caspian Sea legal regime and the convention on the law of the Sea). *Majale-ye Oloum-e Siyasi (Political Science Magazine), 217*, 71.

Hersani, S. (2006). Charchoube Teorike Siyasate Amrica dar Hozeye Daryaye Khazar (Theoretical framework of America's policy in the Caspian region). *Khabargozariye Mehr (Mehr News Agency), 12*, 41.

Kamran, H., Nami, A. M. H., Ja'fari, F., & Mohammadpour, A. (2010). Regim-e Hoqouqi va Tahdid-e Hodoud-e Darya-ye Khazar (The legal regime and restriction of the Caspian Sea Ambit). *Majale-ye Goqrafi (Geography Magazine), 7*(20, 21), 25–46.

Koulaei, E. (1997). Nirouhay-e Pasdar-e Solh-e Rousiye, Bimha va Omidha (Russian peacekeepers, fears, and hopes). *Faslname-ye Motale'at-e Asiyay-e Markazi va Qafqaz (Caucasus and Central Asian Studies), 2*(15), 215–228.

Kumar, P. (2009). The unrealized dream of Caspian Oil. *International Politics, 2*(4), 13.

Mamedov, R. (2001). International legal status of the Caspian Sea: Issues of theory and practice. *Turkish Yearbook of International Relations, 32,* 217–259.

Mehdiyoun, K. (2000). Ownership of oil and gas resources in the Caspian Sea. *The American Journal of International Law, 94*(1), 179–189.

Mirfakhrai, S. (2005). Rejim-e Hoqouqi-ye Daryay-e Khazar az Vagarai ta Hamgarai (Caspian Sea legal regime, from divergence to convergence). *Pajouhesh-e Hoqouq va Siyasat (Law and Politics Studies Studies), 10,* 179–200.

Mojtahed-Zadeh, P. (1999). *Regime Hoqouqi-ye Daryay-e Khazar, Lozoum-e Tabyen-e Manafe-e Melli (The Caspian Sea's legal regime, the necessity of defining national interests).* Tehran: Moasese-ye Ettelaat.

Mojtahed-Zadeh, P. (2009). Didgahha-ye Qeye Karshenasi va Chalesh-e Manafe-e Melli dar Daryay-e Khazar (Non-expert viewpoints on the challenges of national interests in the Caspian Sea). *Bandar va Darya (The Harbour and the Sea), 15,* 5–70.

Navazani, B. (1991). *Ahdname-ye Mavaddat-e Iran va Shoravi (The Soviet-Iran Friendship Treaty).* Tehran: Nashre Hamrah.

Ögütçü, M. (2003). Caspian energy and legal disputes: Prospects for settlement. *Oil, Gas & Energy Law Journal (OGEL), 1*(2). At: https://www.ogel.org/article.asp?key=100. Accessed on Dec 12, 2017.

Omidi, A. (2006). Ijaad-e Mantaqe-ye Moshtarek, Raah-e Halli baray-e Boron Raft az Bonbast-e Rejim-e Hoqouqi-ye Daryay-e Khazar (Creating a communal region as a solution for excluding the deadlock of the Caspian Sea Legal Regime. *Faslname-ye Motale'at-e Asiya-ye Markazi va Qafqaz (Caucasus and Central Studies Journal), IV* (4), 177–194.

Riyahi, M. A. (1990). *Sefarat Nameha-ye Iran, Gozaresh-e Mosaferat va Ma'mouriyathay-e Safiran-e Osmani dar Iran (The Iranian Embassy, letters travel and mission reports of Ottoman Ambassador in Iran).* Tehran: Tous Publications.
Shariat Bagheri, M. J., & Jahani, F. (2001). Tahavolat-e Rejim-e Bahrebardariy az Manaf-e Moshtarek-e Naft va Gaz-e Darya-ye Mazandaran (Operational regime changes of communal oil and gas resources of the Caspian Sea). *Faslname-ye Rahbord (Journal of Rahbord), 20*(58), 147.
Shouri, M. (2007). Naft-o Gaz-e Khazar; Taghabol-e Mnafe' (Oil and gas of the Caspian Sea; The contrast of interests). In *Asian Energy,* 8. http://www.isrjournals.com/fa/special-articles/1083-special-article-farsi53.html. Accessed on Dec 12, 2017).
Torabi, S. (2002). Moshabehatha va Moqayeratha dar Mavaze-e Keshvarhay-e Saheli-ye Khazar (Similarities and Discrepancies in the positions of the Caspian Littoral Countries). *Faslname-ye Motale'at-e Asiya-ye Markazi va Qafqaz (Journal of Central Asia and the Caucasus Studies), 35,* 97–110.

Abdollah Baei Lashaki (Iran) is a Lecturer at Islamic Azad University of Chalous who specialises in the Caucasus and Central Asian regions, as well as in Iranian Foreign Policy. He holds a B.A. in Political Science from the Islamic Azad University of Chalous, and an M.A. in International Relations from the University of Guilan. He has published extensively on Iran's relations with Turkey and the South Caucasus countries, as well as on Russian and American policies and interests in the Caspian Sea.

Masoumeh Rad Goudarzi (Iran) is assistant professor at the Department of Political Science, Faculty of Literature and Humanities, University of Guilan (Iran). She specialises in Islamic Political Thought and Politics and Government in the Middle East and Caucasus. She holds a B.A. in political Science from Shahid Beheshti University, Tehran; an M.A. from Tehran Islamic Azad University; and a Ph.D. in Politics and Government from UPM, Malaysia. She is the author of *Ayatollah Khomeini and the Subject of Islamic Government* (Germany: Lambert Academic Publishing, 2011); has received several research awards, and is published in reputable peer-reviewed journals.

Chapter 5
Energy Resources of the Caspian Sea: The Role of Regional and Trans-regional Powers in Its Legal Regime

Dayyan Janbaaz and Mehrdad Fallah

Abstract The Caspian Sea is the largest body of closed water between Asia and Europe. Since the collapse of the Soviet Union, which brought entirely new conditions, it is also one of the most important regions in the world in terms of geopolitical strategy. Consequently, regional and international competition has intensified, creating regional conflicts and instability accompanied by significant levels of insecurity. Notably, various ambiguities attend the Caspian Sea's legal regime and have thereby complicated efforts to exploit the region's oil and gas resources. Moreover, environmental issues plague expansionist approaches with a lack of cooperative accords among littoral and trans-regional powers, which have contributed to poorly synchronised activities in addition to a failure to properly allot border demarcations. Focusing on the effective components of these challenges, the authors outline regional energy resources and descriptively analyse local and trans-regional influences on the Caspian Sea's legal regime.

Keywords Caspian sea · Legal regime · Geopolitics
Energy resources · Regional and trans-regional powers · Caspian borders

5.1 Introduction

The Caspian Sea in the Eastern Caucasus—Caspian derives from the Caspi nation of the Caspian region—is 1200 km long and 300 km wide (on average) and has been navigated for over 7500 years (Adeebfar 2001: 23). In the late seventeenth century, Iran and Russia became major powers adjacent to the sea and maintained their positions until the twentieth century. Meanwhile, oil extraction began in the late

---

D. Janbaaz (✉)
Hadishahr Unit, Payam-e Nour University, Babolsar, Iran
e-mail: dayan_janbaz@yahoo.com

M. Fallah
University of Mazandaran, Pasdaran St, 4741613534 Babolsar, Mazandaran, Iran
e-mail: mehrdadfalah70@yahoo.com

M. Ebrahimi et al. (eds.), *The Dynamics of Iranian Borders*,
https://doi.org/10.1007/978-3-319-89836-0_5

nineteenth century in Azerbaijan, which initiated regional events leading to today's international disturbances. Of note is that during peak production in 1901, Azerbaijan provided fifty per cent of the world's oil production.

The early 1920s witnessed Russia and Iran's respective consolidations as the region's major powers. Russia gained control of the majority terrestrial sphere, while Iran took over the largest southern sphere of the Caspian region. They were the only countries bordering the sea according to two signed pacts, and this de facto border arrangement was officially recognised until the collapse of the Soviet Union (Karimov et al. 2011: 6). After the Soviet collapse, a totally new global geopolitical situation arose that focused the international community's attention on utilizing the region.

Oil and gas revenues fostered new international competition as ethnic conflicts over power, in addition to terrorism and Islamism, altered the region's security into the 1990s. Hence, the Caspian Sea's legal regime faced neo-border consolidations with remarkable uncertainty. Environmental issues affecting the Caspian Sea were also of great concern. Actions taken to solve the ongoing environmental and biodiversity crisis included initiatives by the Commission of Living Resources of the Caspian Sea, the Convention of Tehran, the Caspian Environment Programme, etc. Pollution and matters concerning biological resources remain unresolved to this day. Given so-called influential factors, this paper attempts to investigate the role of regional and trans-regional powers in the development of the Caspian Sea's legal regime while also addressing its energy resources. Regarding conflicts between the Caspian Sea's littoral nations, the potential value of its energy resources gives much cause for concern and attention, even from other regional and trans-regional players. In sum, the final composition of the Caspian Sea's legal regime and the determination of Iran's share in the matter depend on far more than mere agreements between five countries.

5.2 Estimating the Importance of Energy Carriers of the Caspian Sea

The main economic domain in the Caspian Sea is its energy resource exploitation sector. Energy resources are highly significant for governmental powers, such that the more resources a government has, the more power it can wield. In fact, the geostrategic discourse has become a geo-economic discourse since 1989. The superior role of economic forces in international relations (geopolitical energy) has gained a high position worldwide. The new international political economic structure of energy and control over energy transfer has become the Achilles heel of many industrial regions and countries, including the European Union, China, India, South East Asia and North America. In this regard, geopolitical power now concerns effects from various aspects of energy resources on policy, decision-making, increasing capabilities and different national and governmental relations. Due to increased sharing of petroleum and natural gas in the global energy market, their exploitation now has a unique position in international relations and policies. Considering all the above, the energy carriers in this research concern the established and unexploited oil and gas resources of each coastal region around the Caspian Sea, to include energy transport lines and foreign interests.

5.2.1 *Energy Reserves of the Caspian Sea*

Estimated energy reserves in the Caspian Sea region vary. Differences in estimating energy Reserves adjust regional priorities and can cause disputes between major powers. The US Energy Information Administration estimated the Caspian Sea's oil reserves at 20–40 billion barrels, and gas reserves at 200 trillion ft^3 (EIA 2013). According to experts, these represent four per cent and six per cent respectively, of the current known global supply (Mohammadi 2009: 32).

The US Department of Energy and International Monetary Fund declared that the volume of oil reserves in the Caspian Sea was 178 billion barrels. Another estimate posits three per cent (oil) and three per cent (natural gas) of world production (EIA 2013). In 2010, more than 2.5 million barrels of oil were produced in the Caspian region daily, while natural gas reserves were estimated at 100 billion cubic metres (Karimov et al. 2011: 11). Table 5.1 presents current estimates of the potential reserves.

5.2.2 *Oil and Gas Pipeline Routes*

One issue that differentiates the Caspian enclave from North Sea oil fields and adjacent waterways is the difficulty encountered regarding the export of production to global markets. The Caspian territory is far removed from global consumers. Azerbaijan and Kazakhstan oilfields, for example, are located far from the sea. Hence, domestic producers have difficulty transporting oil to ports servicing international and even domestic markets. Consequently, selecting routes and constructing pipelines is a complex process that occupies major investors and affects international and trans-regional relations.

Increasing numbers of Caspian entities and legal claims have joined with other factors, such as global dependence on hydrocarbon energy reserves, constant Middle East tension, and the re-emergence of Russia as a leading player in twenty-first century energy policy transport to markets. All have focused the attention of huge industrial powers on the region more than ever before (Nosrati and Etaat 2009: 4). Meanwhile, minor coastal countries encircling the Caspian Sea suffer economic poverty and are in urgent need of revenue for the implementation of their development plans. These smaller nations must link with Ocean Communication Networks to enter the international economic system. Nevertheless, during the selection of

Table 5.1 Caspian Basin's Proved and Probable Reserves

Country	Crude oil and lease condensate (billion bbl)	Natural gas (Tef)
Azerbaijan	**8.5**	**51**
Offshore Caspian	6.8	46
Onshore Caspian	1.7	5
Iran	**0.5**	**2**
Offshore Caspian	0.5	1
Onshore Caspian	(s)	1
Kazakhstan	**31.2**	**104**
Offshore Caspian	15.7	36
Onshore Caspian	15.5	68
Russia	**6.1**	**109**
Offshore Caspian	1.6	14
Onshore Caspian	4.5	95
Turkmenistan	**1.9**	**19**
Offshore Caspian	1.1	9
Onshore Caspian	0.8	10
Total Caspian	**48.2**	**292**
Offshore Caspian	19.6	106
Onshore Caspian	28.6	186

Source EIA (26 August 2013)
The bold show total amount of each country reserves

proper transport routes, political advantage waxes more important than economic and technical gains due to the machinations of several regional and international powers (Nosrati and Etaat 2009: 11–12). Consequently, four general routes are currently supported: north/south routes by Iran and Russia; and east/west routes by the US and Europe. Table 5.2 presents an overview of these transport lanes.

Table 5.2 Major Caspian oil and natural gas export routes

Destination	Pipeline	Status	Content	Estimated capacity	Transit route (origin-destination)	Major source fields	Owner
To European markets	Baku-Tbilisi-Ceyhan (BTC)	Operating	Crude oil	1,000,000 bbl/d	Kazakhstan-Azerbaijan-Georgia-Turkey	ACG, ShahDeniz, Tengiz	BTC Pipeline Co
	Caspian Pipeline Consortium (CPC)	Operating	Crude oil	684,000 bbl/d	Kazakhstan-Russia	Tengiz	Transneft, Chevron Caspian Pipeline Consortium, LukArco, Exxon Mobil, Rosneft/ Shell, Agip, Oryx, BG, KazMunai Gas, BP
	Uzen-Atyrau-Samara	Operating	Crude oil	600,000 bbl/d	Kazakhstan-Russia	Tengiz	Transneft
	Baku-Novorossiysk (Northern route export pipeline)	Operating	Crude oil	100,000 bbl/d	Azerbaijan-Russia	Sangachal	Transneft
	Central Asia Centre gas pipeline system (CAC)	Operating	Natural gas	eastern branch: 2200 Bcf western branch: 120 Bcf	Turkmenistan-Uzbekistan-Kazakhstan-Russia	Dauleta-bad	Gazprom, Turkmen-gaz, Uzbekneftegas, KazMunai-Gas
	Baku-Tbilisi-Erzurum (BTE, South Caucasus pipeline)	Operating	Natural gas	280 Bcf	Azerbaijan-Georgia-Turkey	Shah Deniz	BP, Statoil, SOCAR, LUKOIL, Total, Naftiran Intertrade, TPAO
	Kazakhstan Caspian Transportation System (KCTS)	Proposed	Crude oil	Init. 300,000 bbl/d expand to 800,000 bbl/d	China-Azerbaijan	Tengiz	Gazprom, Turkmen-gaz, Uzbektransgaz

(continued)

Table 5.2 (continued)

Destination	Pipeline	Status	Content	Estimated capacity	Transit route (origin-destination)	Major source fields	Owner
To East Asia markets	Kazakhstan-China pipeline	Phase 1 operating Phase 2 planned	Crude oil	Currently: 240,000 bbl/d expand: 400,00 bbl/d	Kazakhstan-China	Tengiz, Zhanazhol	Kazakh-Chinese Pipeline Co (KCPC)
	Turkmenistan-China gas pipeline	Operating	Natural gas	1,400 Bcf	Turkmenistan-Uzbekistan-Kazakhstan-China	South Yolotan (Galkyn-ysh), Karacha-ganak, Tengiz, Kashagan	Intergas Central Asia
To South Asia markets	Iran oil swap	Operating	Crude oil	200,000 bbl/d	Kazakhstan-Azerbaijan-Iran	n/a	n/a
	Turkmenistan-Afghanistan-Pakistan-India Pipeline (TAPI)	Proposed	Natural Gas	1,000 Bcf	Turkmenistan-Afghanistan-Pakistan-India	South Yolotan (Galkyn-ysh), Dauleta-bad	n/a

Source EIA (26 August 2013)

5.3 Energy Policies of Caspian Powers

As mentioned, its hydrocarbon resources increased the Caspian Sea's international profile as a most promising energy tap for major powers, including the US, European Union, China and Turkey.[1] Regional and international competition challenges each power to mitigate otherwise overly ambitious policies and distribute influence more equitably, so the phrase 'Great Game' in the Caspian Sea has come to demonstrate the importance of the region's energy resources (Sander 2006: 4). The following section addresses their strategies.

5.3.1 Caspian Regional Powers

5.3.1.1 Russia and the 'Return Policy'—Russia Is not a Leader but a Major Player

Russia—the closest and most important power in both the Caspian and Caucasus regions—has special interests and is especially sensitive and reactive to any regional changes. The importance of energy to the Russian economy and Russia's role as an exporter of gas and oil invariably influences its foreign policy (Hill 2002: 30). These revenues represent twenty per cent of Russia's GDP, sixty-five per cent of foreign exchange earnings, and at least thirty per cent of Russia's Federal Government income. Accordingly, due to the reduced hydrocarbon resources of Russia's Siberian fields, especially since approximately forty to forty-five per cent of the country's export income was obtained from oil and gas in 2002, the Caspian region is of newfound significance to Russia (Thomas and Shull 1999: 4).[2] Hence, the region is called 'Back Yard' or 'Near Abroad' (Olcott 2006: 375).[3] On the other hand, Russia is an important patron of countries bordering the Caspian Sea, which naturally seek to maximise benefits from the sea's resources. By signing several agreements with neighbouring countries such as Turkmenistan and Azerbaijan, Russia attempted to exploit Caspian resources for its own benefit and increase its share, especially since it is the only power with the ability to exploit the Caspian's hydrocarbon resources.

The importance of the oil and natural gas produced by the smaller Caspian states has led to considerable improvements in their development in recent years, which has a direct bearing on Russia's economic interests in the Caspian Sea, since these

[1] Azerbaijan, Kazakhstan and Turkmenistan are not considered major powers but rather subordinate powers as their policies align with the interests of major powers (US, Russia, European Union and China).

[2] Please note that the greater portion of Russia's hydrocarbon reserves is not in this region, but in the Yarnt and Karlyan Seas. See Thomas and Shull (1999: 4).

[3] Russia's 'near abroad' policy, announced in late 1993, indicated that Russia still pursued interests and policies of the Soviet Union towards Central Asia and the Caspian Sea. Russia's recent economic and political rise returned her to the Game.

countries have rather small populations with less domestic energy demands than Iran and Russia. As a result, improved production and export for most of these countries is ascribed to both East and West, which has brought Russia and its Western allies into competition over strategic planning for oil and gas export routes. Since the three major Caspian energy producers are former republics of the Soviet Union, and since transit infrastructure for pipeline routes was established during the Soviet era, energy transport is a major geopolitical concern to Russia. On the other hand, despite Russia's proximity to the Caspian and Caucasus regions associated with international transport systems, Russia is not regarded as a strategic ally by some Central Asian countries that are reluctant to be dependent on Russia. In recent years, Russia only strengthened such attitudes by cutting gas exports to Georgia and Ukraine.

Recent Russian policy indicates an effort to rebuild its empire by using energy as a weapon. The country emerged as a global player in 2001, especially with regard to natural gas exports (Saunders 2008: 9), and appears to have ambitions towards global hegemony in its new approach by showing increasing influence and energy leveraging. In this regard, Vladimir Putin[4] adopted an aggressive policy towards the recovery of regional hegemony after his election on 26 March 2000 (Ögütçü 2003: 24–25). Since then, many Russians have ridiculed Boris Yeltsin's behaviour towards leaders of foreign countries as 'too intimate' and have also criticised Andrei Kozyrev's submission and reconciliation policy (Legvold 2008: 13–14). Consequently, many analysts opine that Putin's severe tone re-introduces Russia as an influential power in global developments, directly emerging on the international scene with the hefty weight of its energy resources. Accordingly, during Putin's presidency and due to the abatement of foreign investment in the region, Russia focused on two activities in the Caspian Sea region: one, to determine maritime boundaries between Russia and its two neighbouring countries, namely Kazakhstan and Azerbaijan; and two, to gain increasing control over energy export routes (Shouri 2007).

Given these issues, Russia remains in urgent need of hydrocarbon resources to strengthen its economic power while continuing hegemonic competition with the United States. Accordingly, Russia casts a special gaze upon the Caspian while attempting to strengthen its position against its rival (the USA) by further exploiting its regional interests. In this regard, Russia's purposes to strengthen its domination place the interests of the US in stabilizing the region, devoid of Russia's hand as a rival, at substantial risk (Nanay 2001: 487).

Russians believe that the US is trying to create a strip of security from the Mediterranean Sea (to the East) towards southern Russia. They also believe the major factor that will aid Americans success is the region's energy resources (Dehghan 2003: 93). Therefore, although domestic political developments have altered Russia's Caspian policy and placed it more and more in accord with the US, increasing American interventions have caused much anxiety in Russia. Such worry goes beyond NATO expansion and directly concerns Russian policy and several cases of US intervention

[4]Putin has said: 'In the past, the Tsar moved towards conquering the region. Today, we move to gain gas and pipelines.' For this reason, since signing the Energy Charter Treaty in 1991, Russia has yet to confirm it. See Kouzegar Kaleji (2007: 106).

in the Caspian Sea, including (i) American support for the seabed's division; and (ii) the Baku-Ceyhan pipeline as a mass export route from Azerbaijan; and (iii) American cooperation with Azerbaijan and Turkmenistan in settling their disputes over the delimitation of national regions (Ögütçü 2003: 26).

In the case of Europe, experts criticizing European Union policy believe Russia intends to make Europe energy dependent on its Gazprom Company. Another group believes this goal is modest compared to their prediction that Russia means to influence Europe against the United States. This objective arose soon after the Soviet collapse, but the US did not allow Russia to follow it at the time (Koulaei 2008: 77). Accordingly, Russia seeks to prevent energy transport outside Russian controlled routes by different means. Russia signed long-term contracts with Kazakhstan and Turkmenistan to sell their energy resources through Russian transport systems. These contracts were signed during Putin's presidency with the presidents of Kazakhstan and Turkmenistan in May 2007, thereby granting Russia more control over the region's energy transport. In another move, Russia took leadership of an oil export cartel (Petroleum Exporting Countries) that met in Doha in April 2007. Forming the cartel strengthened Russia's control over Caspian energy resources and allowed Russian control over the transport of energy reserves to Europe. A third manoeuvre involved construction projects designed to directly compete with the Nabucco project. These include: 'Blue Stream', which supplies natural gas to the Balkan region, and 'Blue Stream II', which transports natural gas from Russia to Bulgaria through the Black and Baltic Seas, a project that leads directly to Germany (Marketos 2009: 3–4). Therefore, of all powers currently seeking to influence the Caspian region, Russia's energy policy is obviously superior. Russia's strategy seeks to maximise transport routes, control Caspian and Central Asian resources, and optimally exploit the region's oil and gas reserves (Marketos 2009: 3).

Finally, Russia relies on the Caspian region as a vital factor for security and energy concerns. Three facilities enhance these interests: (i) integrating the Commonwealth of Independent States (CIS) as a Russian dominated region by using military, economic and political leverage to *naturalise* its former republics; (ii) international recognition of its peaceful role in assuming exclusive leadership of the CIS; (iii) Russia's presence as the chief guarantor of peace and stability in the region (Terterov et al. 2010: 194). Hence, Russia's policy has thus far been successful in the development of special relations with coastal and trans-regional countries around the Caspian Sea.

5.4 The Islamic Republic of Iran

Iran's policy views Caspian and Caucasus territories as sub-regions while it focuses more attention on the Middle East and northern Africa. However, Iran has experienced a considerable geopolitical upgrade in its Caspian position. Middle East and Persian Gulf tension joined to US inability to solve regional crises served to highlight Iran's role in regional developments (Karimov et al. 2011: 31).

Iran considers its presence and involvement in the Caspian region a kind of securitization of 'Vital National Interests'. From the early twenty-first century, Iran has wrought significant change in its position towards countries standing between Russia and the Persian Gulf. If North Shields once protected Iran against a perceived Soviet threat, a newly created route forms an internationally recognised geopolitical borderline that protects Central Asian countries from the Russian bear. Iran's strategic position is simply defined as finding the shortest, safest and most economic route to transport energy to the high seas and global market (Ahmadi-Lafouraki and Mir-Razavi 2005). Hence, Iran has considerable leverage to negotiate its objectives as a practical means to potent diplomatic relations and regional alliances.

In spite of its advantages and potential, the Islamic Republic of Iran has been marginalised as far as oil and gas resources are concerned among Caspian powers. This is largely because of American opposition to Iran's participation in regional oil and gas projects. The US antagonism is deeply rooted in unresolved post-revolution sentiment, hence America seeks to control and limit Iranian interests. Nevertheless, Iran remains the best route for energy transport, especially, and even traditionally, to East Asian markets. Nevertheless, depriving Iran of its natural interests and geographical facilitations only increases transport costs for regional players, who are then unable to export energy resources via the most economic route (Koulaei 2007: 25). In addition, increasing pressure by western nations, led by the US, seeks to halt Iran's nuclear energy programme, causing even more political and economic sanctions. This very issue led to Iran's rapprochement with Russia as a buffer for the political and economic impasse that positioned it as a shield opposing the presence of aliens in the region, particularly the West.

All in all, instead of utilizing its unequalled geographic potential, Iran became a useful Russian ally of sorts during the era of the post-Soviet collapse. In essence, Iran handed over all areas of competition in favour of Russia's continued confrontation with America and Europe (Koulaei 2010: 85). Iran is also aware that Russia's interests in the region, whether in terms of energy transport or raw exploitation, are contrary to Iran's vital interests. Present interactions between Iran and Russia's 'Near Abroad' are solely based on a kind of tactical passivity. Meanwhile, concern grows over the presence of Western powers, particularly NATO, in the Caucasus region, which, in turn, serves to foster greater compatibility between the foreign policies of Iran and Russia in the region. Thus, Russians view the Islamic Republic of Iran as both a potential rival in energy export to Europe and as a tactical ally against the US and the Commonwealth of Independent States. In other words, Russia has taken a purely tactical position regarding Iran that complements its strategic opposition to the West (Kaczmarski and Gorecki 2009: 1).

As for the Republics of Azerbaijan, Kazakhstan, and Turkmenistan, despite American investments and Western support for Azerbaijan and Kazakhstan as tools to develop their energy interests—not withstanding Turkmenistan neutrality—Iran has attempted to somehow neutralise Western intentions regarding its marginalised role in the region. It has done this by establishing balanced trade relations and adopting a tension-developing policy in relation to the so-called 'Republics' (Bozorgi and Hosseini 2010: 17–18).

As for Iran's relationship with the European Union, the latter seeks new resources for energy transport and the Islamic Republic can surely service such resources. Natural gas reserves in Iran and Turkmenistan can readily enhance existing infrastructure and develop new facilities across the country that would consequently join natural gas transport routes to Europe. Such activity would only increase regional security and greatly benefit regional nations. Naturally, high domestic consumption by energy carriers, especially of natural gas, and the obvious failure to project a serious policy towards increasing exports, caused Europeans to hesitate over Iran's dubious potential to export natural gas, giving the European Union a reason to allocate only a small portion for any Iranian role.

An example is the Nabucco natural gas pipeline, which supplies at least 15 billion cubic metres of natural gas from Iraq annually. Iran's portion was estimated as equal to current exports to Turkey, although Iran's natural gas reserves are not comparable to those of Iraq. On the other hand, Iran's joining with independent Caspian countries to autonomously export natural gas to Europe contravenes Russian policy, which opposes direct relations of these regions with Iran (Shadivand 2013: 40). Moreover, the West's bias against Iran, as led by the US, aims to remove Iran from the global equation, particularly energy markets. Therefore, Iran's passive position only augments global sanctions. One can only imagine that a wiser decision on the part of Iran, accompanied by active diplomacy and specific programmes in international energy relations (oil and gas) would better secure the country's long-term interests.

5.4.1 Trans-Regional Powers

5.4.1.1 The United States and Strategic Surprise

After the Soviet collapse, the US made massive efforts to shape a unipolar world system or New World Order. With this specifically in mind, America considered the Caspian Sea and Middle East major suppliers of world energy for the twenty-first century.

Consequently, United States foreign policy for the Caucasus and Caspian Sea has undergone different stages, the first of which extended from 1991 to 1993. A major American concern was (is) the influence of Iranian Islamism and spread of fundamentalism in the region. During those years, due to significant events in Russia, the US prioritised Russia and the remaining republics of the Soviet Union, with Central Asia and the Caucasus republics holding lesser importance. At this stage, limiting Iranian and Russian relations was a top priority in line with America's well-defined international responsibility as the underwriter of Western security by accessing energy resources as its main purpose for such international interventions. Washington abandoned former policies, including the trivializing of energy-based resolves in lieu of other issues such as the emergence of minor economic tensions

(Oktava 2005: 17).[5] After the emergence of the second Persian Gulf crisis in the early 1990s (invasion of Kuwait by Iraq), access to oil and gas resources outside the Middle East and Persian Gulf became more important to the United States. Although Caspian Sea resources are by no means comparable to Persian Gulf reserves, they were/are considered a suitable alternative in times of crisis.

The second foreign policy stage followed from 1994 when American oil companies gradually became active in regional oil projects as oil gained significant currency among US foreign policy in the Caspian region. Simultaneously, America pursued other political moves such as its intervention in the Karabakh conflict. A third stage followed the events of September 11 and occupation of Iraq with access to its oil resources. Hence, the importance of Caspian oil declined somewhat for America. However, interest in the region regained focus with sharp rises in the price of oil (Koulaei 2008: 50). America's special geopolitical view towards the Caspian Sea as a main supplier of global energy requirements for the twenty-first century then reappeared with renewed vigour. Hence, American efforts to overshadow the Caspian region and gain control of the world have garnered much international attention.

Currently, this issue grows in importance for America, mostly because of oil and the region's geopolitical potential for diversified energy transport systems and consequent security issues (Shouri 2007).[6] Jay Main raised concern over US policies towards Russia that appear to confirm America's regional position with questions such as: 'If Russia comments on the Gulf of Mexico's crucial and strategic interests, what will the United States' reaction be?'—or—'If there weren't oil and gas reserves in the Caspian Sea, would the US be willing to enter this field?' (Main 2005: 15). Answers are found in statements by Richard Morningstar:

> … and we believe that local borders are finally the best means for the application of influence. It is important that solving these issues would not prevent the extraction and development of resources in the Caspian Sea (Ögütçü 2003: 26–27).

Thus, overall US strategy in the Caspian region aims to control the region's energy resources, which has not only resulted in influence and political domination by the US over these so-called countries, but has also helped America plan aggressive programmes against China, Russia and Iran for the near future (Balmasov 2011: 22–23).

The US and its allies posit that their proposed pipeline route would diversify energy supply resources and strengthen regional independence. However, it is actually intended to upstage Iranian and Russian influence (Ögütçü 2003: 17–18). From

---

[5]Zeynep Oktava believes that since 1990 the US government has prioritized the improvement of America's energy-based projects so that the energy transport rate would not only continue to American markets, but also to its economic partners. See Oktava (2005: 17).

[6]Richard Morningstar, Bill Clinton's supreme adviser in the Caspian oil and gas issues, indicated the main elements of this policy: "… newly developed Caspian countries are located in a region surrounded by different problems and risks. These countries that are captive between Russia in Southern and Eastern Iran … have found opportunity, for the first time, to gain control over their destiny. To achieve this goal, as well as strengthen our own security in accessing energy, the region's independence to access its own energy provides opportunities for our company and will be an absolute benefit for the United States." See Shouri (2007).

the Western perspective, a major risk is the proposed liquid natural gas pipeline that joins Russia to the Caspian Sea and Middle East, and continues eastward to India and China. This alone can lead to constantly increasing gas and oil import costs for America and Europe, while inclining the strategic balance in energy security towards gas-exporting countries led by Moscow (Pabst 2009: 175–176).

The US also seeks to encourage Caspian countries to exploit their resources, particularly oil, so they can improve their economic conditions. Such a change can improve regional conditions and provide a more suitable consumer market for American companies, given that American economic prosperity, as one of the largest economies in the world, is also dependent on increasing growth via the economic convergence of the world's major economic hubs. Hence, America seeks to exploit extant resources along with new energy sources to stabilise these economic hubs (American and European systems). Direct influences by US authorities on various contracts with Caspian countries, along with intervention in the Caspian Oil Consortium while encouraging coastal countries to join NATO, in addition to actually menacing working oil companies in the region, are examples of America's political invention in economic development programmes in Caspian republics.

America also plays a significant role in Europe's strategy towards Central Asia, the Caspian region, and Black Sea development. Here the US supports the European Union by concentrating on opposition to Russian energy transport programmes through the Caspian Sea to Europe. However, unlike European efforts to establish permanent and balanced relations with Russia, the US never hesitates to confront Russia. In addition, America continually amplifies any conflict between Russia and its neighbours.

US policies towards China are related and of primary importance, as 2009 marked an American milestone regarding China's Central Asian policies. As the initiation of the Turkmenistan pipeline route to China approached, US opposition increased and the Senate Foreign Relations Committee met to discuss the issue in February 2009. For the first time, Morningstar emphatically announced China as a serious rival to America's Central Asian energy policies. He stated "China will gain increasingly significant influence in Central Asia … it is difficult for us to compete with China in some of these countries" (Boroujerdi et al. 2011: 21). Hence, given Russian and Chinese proximity, a clash of regional interests is probable, although global conditions, policies and infrastructure challenges encompass regional interactions between both nations. Given all the above, two general statements sum up American policy and national interests: (i) secure access to non-Middle Eastern oil and gas reserves; (ii) ensure American influence in Russia's backyard (Main 2005: 14).

5.5 European Union and Western Focus: Policy Matters

Since global energy concerns hold the utmost strategic significance for the twenty-first century, geopolitically, energy markets and circumstances have undergone enormous changes that hold possibilities for both a 'New Cold War' and 'Hot Resources

War'. Entering the fray is the European Union, whose members have precious little of the world's crude oil and natural gas. Their domestic reserves decreased from 1.8 and 4.6% to 0.4 and 0.9%, respectively, in 2011. Decreases in European Union crude oil and natural gas production are considerable. From 1999, the Union's crude oil and natural gas production decreased from 3684,000 barrels/d and 227 billion cubic metres, respectively, to 1692,000 barrels per day and 155 billion cubic metres in 2011 (BP Statistical Review of World Energy June 2013). Currently, fifty per cent of the Union's energy consumption is supplied by imports. Accordingly, given the Union's short- and long-term views, it will import ninety per cent of its crude oil needs and seventy per cent of its natural gas requirements by 2020 (De Jong 2008: 16). The greater portion of these currently derives from the Persian Gulf, Russia and northern Africa. However, given the instability and distance of Middle Eastern energy resources, and the gradual collapse of its domestic production, the Europe Union has increased its focus on the Caspian region. In addition, membership options offered to Eastern European countries for the purpose of energy security have transformed future outlooks for Caspian and Black Sea nations.

Given these issues, geopolitical factors regarding international energy transport security have caused the European Union much concern over the status of other countries. Through its vast regional networks of cooperation for export, transport and import, the Union manages immense influence on a wide range of economic continuity with a view towards the provision of interdependence with its partners in and around Caspian regions. Hence, the European Union views energy security on a vast and comprehensive scale and holds forth that economic and political measures—with the emphasis on new infrastructure including terminals for receiving liquid natural gas (LNG), and a new pipeline route from the Caspian—are initial steps to secure energy resources. By emphasizing its focus on economic and geostrategic issues and long-term energy policies, the Union shows much interest in cooperating with nations of the Caspian region (Shadivand 2013: 37–38).[7]

In this regard, European companies hold considerable stock in existing oil and gas export routes. Currently, six major pipeline routes for crude oil, and four for natural gas actively transport and export Caspian Sea oil and gas to global markets. Among these routes, five crude oil and two natural gas routes service European markets directly or indirectly. Nevertheless, fifty-seven per cent of the capacity for crude oil pipelines, and forty-one per cent of the capacity for natural gas transport remain under Russian control (Shadivand 2013: 38). Hence, several European companies, such as Total, Exxon, and NE, are important shareholders in several Caspian Sea projects and pipelines. For example, European companies like BG, NE, and Shell, together with the American companies Chevron and Exxon Mobil, own 30.2 per cent of Caspian Consortium pipeline shares. Others like BP, NE, StatoilHydro and Total, together with the American companies Chevron and ConocoPhillips, own 60.2 per

[7]European oil experts believe the European Union should take different measures to support and develop hydrocarbon resources in the Caspian Sea while improving access. However, implementation faces serious challenges from rival countries, particularly China and Russia. Nevertheless, the Union's positive position is that it has the financial and technological ability to develop these resources so as to benefit local economies in the region (Shadivand 2013: 37).

cent of Baku-Tbilisi-Ceyhan pipeline shares. BP, StatoilHydro and Total also own sixty-one per cent of the Baku-Tbilisi-Erzurum natural gas pipeline (Shadivand 2013: 38). In other words, the Caspian region is in a good position to securely supply oil and gas and reduce dependency on Russia through the creation of an Asian-Eurasian transport corridor.

Furthermore, the European Union is the world's largest oil and gas importer, requiring approximately twenty-nine per cent of total world energy production. As such, it is strongly influenced by changes in the energy market. Meanwhile, increasing oil prices and the emergence of new players creates an obvious need for new energy security and foreign policies. Because of its energy dependence on Russia and the political instability and insecurity of Persian Gulf countries, the Union continually seeks new policies to better establish stable energy supplies for the future. In this regard, Europe's long-term view holds Caspian natural gas resources in higher regard than its oil resources because the Union is far more willing to utilise gas for its future industrial developments and production (Boroujerdi et al. 2011: 7).

The problems caused by the collapse of the Soviet Union and the emergence of independent specific security and economic policies of Central Asian and Caspian republics have also increased the European Union's interest. The Union positively proposed a policy of non-interference in the internal affairs of these countries in the early autumn. In 2006, Europe's strategy for Central Asia and the Caspian region took form. Activities included cooperation, excepting Turkmenistan, and partnership agreements for ground transportation, energy, internal corporate policies, environmental matters, drug-trafficking interdiction and the fight against poverty (Scaroni 2006: 15). Soon there followed border management issues with special measures regarding the reform of security structures (Scaroni 2006: 34), along with bilateral cooperation contracts for technological aid for the transport of oil and gas to Europe (Boroujerdi et al. 2011: 17–18).

The supply of energy to Europe from the Caspian geopolitical basin also engages global Powers including the United States, Russia, China and Iran. In this regard, Europe's energy dependence on Russia is a major concern. Approximately 77.3 per cent of minerals and raw materials imported from Europe through Russia are supplied to the European Union. Also, a significant portion—forty per cent—of the electricity produced its gas imports from Russia goes to Europe (Marketos 2009: 9–10). Hence, the EU is trying to diversify import sourcing from the Russian and Caspian regions while incorporating energy considerations (Boroujerdi et al. 2011: 9).[8] Therefore, although the ability to replace OPEC with Caspian resources in Europe is low, EU energy imports from OPEC would act as supplements. In fact, European countries do not wish to import energy and tend to limit imports from Central Asia and the Caspian Sea. Plans for the Inogate, Trasca and Nabucco pipelines required

[8]In November 2000, the European Commission edited Europe's strategies for energy supply security in the 'Green Book'. This document examines defects in the Union's energy supply structure, particularly challenges in the coming decade. Consequently, the European Union Commission based its programme on secure energy supplies by diversifying resources, thereby completing the European Union's energy policy. See Boroujerdi et al. (2011: 9).

the creation of charters that offered 'Eastern Partnership'[9] to Caspian and South Caucasus nations, in an attempt to gain more influence in the region. As a result, the Union risked antagonizing Russian foreign policy in the region. In turn, these same countries began cooperating with Europe to establish closer relations. An energy alliance with Azerbaijan was struck in mid-November 2006. Azerbaijan is a plasticiser of oil production for Europe and its geographical position stands as a transit route to Kazakhstan, which is of great importance. Despite this, participation and political interventions by the European Union in the Caspian region are far weaker than those of the United States. However, Kazakhstan and Turkmenistan do not have diplomatic relations with the EU. Azerbaijan is relatively closer to the Union and partakes in the Eastern participation programme and other Western institutions but maintains its political independence. Generally, the European Union's strategies in the Caspian region affected alliances with Russia and the Caspian region that served to diversify its investment in energy resources and energy security. In this regard, the Union's relations with the US, Russia's rival, can affect its policies.

5.6 China and the New Deterrents

China is an important neighbour of the Central Asian republics, one that surpasses Russia in terms of population and economy. After the collapse of the Soviet Union, for political, economic, and security reasons, China tried to gain influence in the region. Since China needs the region's gas and oil, supplying part of China's energy requirements from Caspian countries has become an important issue (Boroujerdi et al. 2011: 421; Ramakant 2006: 147),[10] especially since China is the

[9]European Neighbourhood policy is intended, a policy proposed in 2003 when the European Union was extending eastwards. New 2004 members were followed by Bulgaria and Romania in 2008, which caused the Union to share challenges inherited from the Communist era. As a result, some felt that the Union required a set of policies to meet these issues; one that was not limited to its neighbours but included other countries as well. Their logic was based on a continuum for development: '*as the cooperation between the countries increases and the scope of peace, economic development, and democracy expands, the European Union enjoys more security and stability*'. The initial proposal included features such as the right of eligible countries to benefit from participation in the EU's domestic market with more convergence with these countries, including the free movement of citizens, goods, service and money, in addition to cooperation in energy, transportation and the fight against terrorism. The final policy included other features, such as effective political cooperation and negotiation, supporting for membership in the World Trade Organization (WTO), cooperation in legal immigration for citizens from eligible countries, cooperation against mutual security threats, cooperation in crisis management, improved cultural cooperation and mutual understanding, increased convergence and cooperation in research, mutual war against drug trafficking and organized crime, improved educational, scientific and technological cooperation, and cooperation in security and environmental protection.

[10]A majority of energy experts believe that China's unlimited consumption will cause a crisis that will delimit the international energy discourse. See Ramakant (2006: 147). China, which is close to a new increase in the field of energy, will increase its natural gas consumption from the present three per cent to ten per cent by 2020. Its gas consumption in 2005 was 77.8 billion cubic metres

second highest consumer country in the world due to the rapid development of its industrial sector.

To understand China's approach to Central Asia and the Caspian Sea, in the early 1990s Deng Xiaoping said, "China should remain silent and wait for an opportunity for decision-making" (Mihalka 2007: 34). Hence, Chinese policy towards Caspian energy resources included peaceful relationships while suppressing Islamic movements and repressing secessionist tendencies. China also sought to circumvent the US by limiting its influence in the region in order to secure its own economic and energy interests. In this sense, the Chinese diligently followed pipeline developments from the Caspian Sea to China to prevent them from passing through/to international seas, especially since the Indian Ocean is monitored and controlled by the United States. In other words, a range of security issues associated with the search for oil security, as well as related economic and political issues, is of the utmost concern for China (Andrews-Speed AND Vinogradov 2000: 397).[11] Accordingly, since 1996 China has focused on developing strong economic and military cooperation with Russia to better enable both countries to withstand North America hegemony (Goldstein 2001: 846–847). Nevertheless, China tries to overshadow Russian interests in the Caspian Sea while avoiding political and military roles in the region. Even so, Moscow fully embraces China's efforts to align its policies with Russia (Olcott2006: 375). Their cooperation in the same venue serves to counterbalance American influence in the Caspian Sea; something neither could do on their own (Mihalka 2007: 34). Clearly, Russia's bowing to China's support forestalls US dealings with major players in the region and is of major concern in the west. Furthermore, both China and Russia have always been concerned about NATO's development in the region. Subsequently, they have found mutual ground to combat NATO's security measures and fight against terrorism via the SCO (Shanghai Cooperation Organization) and by signing bilateral agreements with Caspian countries. In general, China's increasing demand for energy places Russia as its strategic ally in the region.

China also considers Iran an economic partner. Both nations share more geopolitical interests than energy and economic concerns. Naturally however, Beijing has to align its relations with Iran in accord with US policy as China's major economic partner, to assure the US of its 'balanced' approach to Iran (Ahmadi and Jalil-Piran 2010: 137).

Having favourable relations with Caspian republics is another factor in China's desire for increasing power. China has very close relationships accompanied by financial support for oil and gas pipeline routes in these countries. China's policy of

while its domestic production was 77.5 billion cubic metres. Hence, China was unconcerned about the import of natural gas. However, the country will soon encounter a deficit of between 70 to 110 billion cubic metres for domestic consumption. Its entire gas imports, including LNG, increased to 18 million tons per year in 2010. Based on predictions, China's need for oil imports will reach 250 million tons annually by 2020 with a recently admitted thirty per cent increase in demand for global oil supplies between 2000 and 2004. See Boroujerdi et al. (2011: 21).

[11]Although the scheme to build a pipeline by China in the Caspian region is incompatible with western designs, it supports economic energy security for China, even if such a pipeline supplies five per cent of China's demand for oil.

'non-interference in their internal policies' has increased enthusiasm in the region for its economic role, which provides significant lifelong benefits for the region. However, the Chinese government also seeks military cooperation with Caspian republics through organizations such as Shanghai, with a particular view towards combating terrorism and racism (Karimov et al. 2011: 23–24). Contrary to the Western perspective, Caspian republics not only welcomed China's momentum but have also entertained difficult negotiations with the West that fiercely constrained the latter's licensing privileges. Furthermore, China's presence in these new dependent countries is not limited to oil and gas interests, but also plays host to a wide range of extremely effective, unrelated economic aspects.

Generally, China's energy diplomacy in the Caspian region is influenced by several factors. First is the attempt to stop dependence on Persian Gulf oil. Secondly, China reaches out to oil producers who are less affected by US policies in order to minimise US influence on any transfer of energy products to its mainland. Thirdly, China seeks to achieve bilateral relationships with Caspian republics for mutual 'self-care' due to fluctuations in production. Finally, a significant component of Caspian energy production is funded by the Chinese government, which encourages Chinese companies to invest in the exploitation of new oil and gas reserves (Murphy 2003: 13).

5.7 Turkey and Its Dichotomous Regional Foreign Policy

Turkey's aspirations towards the Caspian region directly concern energy security as the most valued national interest. Accordingly, Turkey's foreign policy '*minimises problems with neighbouring countries ... in accordance with regional circumstances*' (Aydin 2010: 81). Hence, Turkey has paid keen attention to regional issues since 2006. The nation's foreign policy is termed 'strategic depth'. Its geopolitical basis is that Turkey stands between two gigantic geo-cultural groups; the Islamic Middle East and the secular-progressive West (Terterov et al. 2010: 197–198). In fact, Erdogan's activities in both camps have hinted that Turkey is the superior authority and should, consequently, be elevated in status and influence regarding regional geopolitical energy matters (Marketos 2009: 8-9). Hence, Turkey's energy strategy has been to cooperate with the US and grant considerable collateral movement with Russia regarding energy transport and pipeline routing across the Black Sea, in addition to a wide range of economic operations (Hacisalihoglu 2008: 1867–1872). Although Turkey's becoming a 'regional energy hub' presents challenges between its own and Russia's approach to the West regarding the establishment of pipeline routes (Koulaei 2008: 64),[12] Turkey cooperates with both Russia and America in order to maintain

[12]An example is the problem Turkey has with the Kurds and the attempts by this country to pass a pipeline through that region to the Mediterranean Sea, intended to strengthen its conventional route through the Port of Novorossiysk, which raised serious political conflict with Russia. Each country accused the other of interfering in its domestic affairs. Turkey stands accused of intervening in Chechnyan affairs and Russia stands accused of intervening in Kurdish affairs. See Koulaei (2008: 64). Also, in 2011, Moscow became concerned about coordinated policies between Ankara and

a balance. Hence, building pipeline routes along the southern line of 743 miles at a cost of 3.2 billion dollars not only secured Russia's influence in Turkey, but also assured Russia's export of gas to Turkey (Marketos 2009: 8–9).

As for relations between Turkey and the US regarding the Caspian region, Turkey has always supported US interests in the region. For example, during its participation in the construction of the Baku–Ceyhan route, the US government sponsored Turkey, both politically and economically, to achieve this great regional success (Cornell 1998: 51). Later that year, Turkey gradually began initiating projects focused on its general policy in the Caspian region (Ögütçü 2003: 29). On the other hand, Turkey still needs to improve relations with Russia because sixty-five per cent of Turkey's natural gas consumption is provided by Russia (RAS). Hence, Turkey's participation in pipeline routes supplying Western interests in the region fostered new tensions with Russia over energy transport from the Caspian Sea to the West (Koulaei 2010: 82).

Turkey is also an important strategic partner for the European Union, and its importance to the EU is increasing. Hence, European Union membership opens a new perspective on Turkey's relationship with its Caspian partners (Shilibekova 2008: 57). Turkey is the major oil and gas transit route from the Caspian region and Central Asia into Europe. In addition to Caspian-Euro pipeline routes, Turkey transports Caspian oil and gas to Europe through the Gulf of Ceyhan via Istanbul's airport, furthering support for its membership in the EU (Kinnander 2010: 19–20). Turkey has also diversified its energy supply lines through active participation in the construction of the Nabucco pipeline and the Baku-Erzurum-Ceyhan transport of Shah Deniz gas to Georgia, Turkey and Europe, consequently acquiring an important position in European energy security (Koulaei 2010: 82). Generally speaking, although Turkey actively cooperates with the European Union to improve its position and membership, such cooperation has always been considered Turkey's 'Achilles Heel' in its relations with powerful regional rival countries such as Russia and Iran.

These regional and trans-regional policies, as adopted by greater powers, directly focus on relationships between the Caspian's legal regime and the special interests of the powers that be. Thus, each authority holds a slightly different outlook on the Caspian Sea, which has brought some degree of ambiguity to the Caspian Sea's legal regime discussed later in this study.

5.8 Geopolitical Competition and the Caspian Legal Regime

The study of this Legal Regime exemplifies international geopolitical security interests that are affected by competition for economic and energy interests. In this context, although no agreement exists on the Caspian Sea's legal structure regarding the exploitation of its resources, regional powers have nevertheless actively entered into

---

the West in the Caspian region, as well as in the Middle East and North Africa. Furthermore, both countries hold different positions on Syria, another source of friction affecting bilateral relations.

transactions with foreign companies. Considering the differing views over maritime boundaries held by each of the five regional nations, unilateral activities have necessarily caused tension. Several concessions have been granted to international companies for the extraction of Caspian oil and gas, particularly by new Caspian countries. Resulting conflicts over undivided territory bear the following three decision-making patterns:

1. *Iran-Turkmenistan*: A proposal of international tender (Summer 1997) after negotiations between Iran and Turkmenistan's maritime boundaries in an area shared by both countries. The Turkmenistan government excluded three out of eleven blocks from the tendered proposal in October 1997. Meanwhile, both countries decided to exploit these regions until the Caspian legal regime was determined.
2. *Russia-Kazakhstan*: After Kazakhstan objected to Russian oil operations in the north Caspian Sea, both countries solved the conflict by dividing the region between them.
3. *Azerbaijan-Turkmenistan*: Regional conflict between Azerbaijan and Turkmenistan over the Azeri and Chyrak oilfields ceased as the result of an agreement between Azerbaijan and two Russian companies, LUKOIL and Rosneft, in July 1997. However, Turkmenistan objected and the agreement was cancelled by Russia. Nevertheless, Turkmenistan gained currency by proposing a tender for the same region in 1998. Presently, 'talks' notwithstanding, this conflict remains unresolved.

Caspian Sea energy resources and economic interests arising from its geopolitical strategic position have raised several questions concerning legal protocols that include the following:

1. Is this body of water a sea or lake?
2. How should its domain be divided between its littoral countries?
3. Should its resources be used mutually or individually?
4. Should rules for pipeline constructions passing through the sea be agreed on by all five countries, or only by those countries through which the pipeline passes?

Thus far, answers from the region's littoral countries demonstrate a natural proclivity towards independent gain. Hence, conflicts engage ambiguity over the Caspian Sea's near absent legal regime, which, thus far, has led to unbalanced distributions of oil and natural gas reserves contained in both the seabed and the subsoil (Ögütçü 2003: 39). The attending coastal countries have taken different positions regarding the legal regime in concert with exogenous efforts to access these resources, inevitably affecting the overall rate of progression towards achieving legal solutions.

Given these issues, Western powers, including the US, by encouraging Azerbaijan, Kazakhstan, and Turkmenistan to share the seabed, are attempting to gain control over the transport and production of gas and oil from Iran and Russia to favour their own economic interests. They have also attempted to intervene in the Caspian region's legal regime to enhance their energy security and strategic objectives. The US supports a legal regime based on a comprehensive division of the Caspian Sea that results in a shared boundary for Azerbaijan and Turkmenistan. Consequently, the border link between Iran and Russia would disappear. Accordingly, and without interventions by Tehran or Moscow, Washington could consolidate its strategies in Turkmenistan, Kazakhstan, and Azerbaijan. On the other hand, by pre-empting any formal determination of a legal regime, opportunities for further exploitation are created by the excuse of fighting against Islamic fundamentalism (terrorism).

Reacting to this deadlock strategy, the Russian government proposes an approach in which each country has exclusive jurisdiction over oil reserves within 45 miles of its own borders. This action favours exploration and exploitation by Russian oil companies (LUKOIL and Gazprom), in international regions and those areas adjacent to the oil and gas reserves of Azerbaijan and Kazakhstan, in addition to closing possible bilateral contracts with Kazakhstan. Consequently, the northern Caspian Sea's legal regime, cited earlier, would expire. The action is directed against US efforts to achieve its own strategic objectives. In subsequent years, Russia confirmed Azerbaijan's and Kazakhstan's position regarding the necessity of dividing the Caspian Sea by first participating in regional oil programmes, for example, the cooperation between LUKOIL and an Azerbaijan company to extend Caspian oil resources, and the determination to build a pipeline reaching the Port of Novorossiysk (near the Black Sea) from the Caspian region in Kazakhstan. Another action was Russia's signing of bilateral contracts with Azerbaijan and Kazakhstan to divide the Caspian Sea for mutual use (including shipping, fishing and environmental activities). As a result, Russia created a framework confirmed by the US and its newly dependent coastal countries. As such, it left little opportunity for Iran to determine any component of the legal regime. Moreover, since 1998, large oil resources in the region claimed by Kazakhstan, particularly Kashagan, have been confirmed as the largest reserves and are comparable to those in the Persian Gulf. Hence, Russia took immediate action to determine the maritime borders between both countries. In 2002, Putin signed an agreement with Nazarbayev for equal exploitation (fifty-fifty) of these resources, regardless of location toward the midline (Haji-Yousefi 2005: 133; Shouri 2007).[13]

The Islamic Republic of Iran opposes bilateral agreements between countries of the Caspian Sea and the West directly related to Caspian energy interests. The reason for this opposition is the absence of a specific legal regime. Such bilateral agreements provide these countries with unrestrained opportunities to exploit Caspian resources, thereby leaving less for future Iranian generations.[14] Already, practical disagreements exist over oilfields in Araz, Arav, and Shargh between Iran and Azerbaijan. Similarly, in 2001, an Iranian warship supported by the Air Force forced two Azerbaijani reconnaissance aircraft (Geofizik-3 and Alif Gadzhiev), working under a British Oil Consortium, to dump their discoveries over the Alov (Alborz) field (Laruelle and payrouse 2009: 21).

[13]Russia's attempt to define the Caspian Sea as a 'closed lake' was evaluated. By proclaiming the Caspian Sea a closed lake, Russia tried to prevent the signing of oil contracts with the West by coastal countries without considering Russia's interests and managed to establish its hegemony across the region yet again. See Haji-Yousefi (2005: 133).

[14]The Islamic Republic of Iran hasn't received any proceeds from Caspian oil and gas resources so far.

Thus, the presence of the US and its allies in the region influences some countries to form regional organizations and bilateral cooperation pacts, which, in turn, cause ambiguity towards Caspian borders that present potential risks to Russia and the Islamic Republic of Iran (Main 2005: 16). It has also inspired a political melee between Russia, the US, regional and trans-regional allies regarding access to and agreements on the southern part of the Caspian Sea, thereby providing an even greater incentive to determine borders between Iran (Russia's Ally), Azerbaijan and Turkmenistan (US allies) at the peak of the triangle.

5.9 Conclusion

With regard to the Caspian Sea, this study investigated the energy policies of regional and trans-regional powers. The implications are that regional energy resources and consequent changes have great effects on the strategic orientation of involved trans-regional powers. After reviewing the legal regime, we can conclude that negotiations between Caspian coastal countries to determine maritime boundaries have lasted for over a decade, with many investigators suggesting that mutual solutions do not exist to resolve existing conflicts. Nevertheless, trans-regional countries have exploited these differences to postpone the formulation of a Caspian Sea legal regime in lieu of achieving their own goals. Therefore, extant resources in the Caspian Sea pose a double-edged sword. They can be a source of income and means of cooperation for neighbouring countries, or cause potential risks (Iran and Russia) due to an undetermined legal regime and the presence of major powers in the region.

Thus, as Critchell believes (Critchell 2008), no single country plays a key role in the Caspian Sea. Furthermore, any proposed solution from countries outside the region is a great barrier to any determination of a legal regime and free exploitation of its resources. As a result, because regional countries are unable to reach an agreement based on international law and former treaties, present conflicts on how to divide the Caspian Sea and its resources will continue. Hence, it is not surprising that continual disagreements over maritime boundaries encourage the countries involved to prepare for military defence.

References

Adeebfar, T. (2001). *Geopolitical dimensions of the main export pipeline in the Caspian region: The Baku-Tibiis-Ceyhan pipeline and the events of 11 September 2001*. Canterbury: University of Kent.

Ahmadi-Lafouraki, B., Mir-Razavi, F., (2005). *Rahnamay-e Mantaghe-ye Khazar va Keshvarhay-e Asiay-e Markazi va Ghafghaz-e Jounobi (The guidebook to the Caspian Region and Countries of Central Asia and South Caucasus)*. Tehran: Abrar Mo'aser Press.

Ahmadi, H., Jalil-Piran, D. (2010). Rezhim-e Amniyat-e Daryay-e Khazar pas az Foropashi-ye Ettehad-e Shouravi ta 2009 Mavane' va Chaleshha (The security regime of the Caspian Sea from the collapse of the Soviet Union to 2009: obstacles and challenges)", in: Faslname-ye Tahqiqat-e Siyasi va Bynalmelaliy-e Daneshgah-e Azad-e Islami-ye Shahreza *(The Political and International Research Journal of Azad University of Shahreza Unit), 3*, 137.

Andrews-Speed, P., & Vinogradov, S. (2000). China's involvement in Central Asian petroleum: Convergent or divergent interests? *Asian Survey, 40*(2), 377–397.

Aydin, M. (2010). Turkey's caucasus policies, UNISCI Discussion Papers, No. 23, 77–91.

Balmasov, S. (2011). The United States and the Caspian Sea Countries are chummy: But against whom? *New Eastern Outlook*, 22–23.

Boroujerdi, A., Ansari-Zadeh, S., & Karami, M. (2011). Zheoeconomy-e Darya-ye Khazar va Ta'sir-e an bar Amniat-e Enerzhi-ye Etehadiye-ye Ouropa (The geoeconomy of the Caspian Sea and its impact on the energy security of the European Union). *Faslname-ye Rahbord (Journal of Rahbord), 20*(60), 7.

Bozorgi, V., Hosseini, M. A. (2010). Sazman-e Hamkari-ye Shanghai: Tahavolat-e gozashte va Cheshmandaz-e ayande (Shanghai cooperation organization: The past developments and future landscape). Motale'at-e Orasiya-ye Markazi *(The Journal of Central Eurasia Studies, 3*(7), 17–18.

BP Statistical Review of World Energy June 2013. At http://www.bp.com/content/dam/bp/pdf/statistical-review/statistical_review_of_world_energy_2013.pdf. (August 22, 2013).

Cornell, S. (1998). Turkey and the conflict in Nagorno Karabak: A delicate balance. *Middle Eastern, 34*(1), 51–72.

Critchell, K. (2008). The law and politics of the Caspian Sea. *Journal of International Affairs, 61*(2), 298–299.

De Jong, J. (2008). *The third eu energy market package: Are we singing the right song*. Briefing Paper 8, Clingendael International Energy Programme.

Dehghan, F. (2003). *Barrasi-ye Manabe'-e Naft va Gaz-e Houze-ye Daryay-e Khazar va Tasir-e an bar Amniat-e Melli-ye Jomhouri-ye Islami-ye Iran (An examination of the oil and gas resources of the Caspian Sea and their impact on the national security of the Islamic Republic of Iran)*. Tehran: Hame Press.

EIA. (2013). Caspian Sea region: Overview of oil and natural gas in the Caspian Sea region (2013). Online at http://www.eia.doe.gov. Accessed November 28, 2017.

Goldstein, A. (2001). The diplomatic face of China's grand strategy: A rising power's emerging choice. *The China Quarterly, 168*, 835–864.

Hacisalihoglu, B. (2008). Turkey's natural gas policy. *Energy Policy, 36*(6), 1867–1872.

Haji-Yousefi, A. M. (2005). *Siasat-e Khareji-ye Jomhouri-ye Islami-ye Iran dar Partov-e Tahavolat-e Mantaghei; 1991–2001 (The foreign policy of the Islamic Republic of Iran in the light of regional developments; 1991–2001)*. Tehran: Daftar-e Motale'at-e Siyasi va Bynalmelali.

Hill, F. (2002). Russia: The 21st century's energy superpower? *The Brookings Review, 20*(2), 28.

Kaczmarski, M. E. P., & Gorecki, W. (2009). Iran's position in Russia's foreign policy and Russia-American relations. *CES Commentary, 15*(4), 1.

Karimov, R., Chausovsky, E., & Makili-Aliyev, K. (2011). *The Caspian Basin: Geopolitical and the future balance of power*. Baku: SAM Center for Strategic Studies.

Kinnander, E. (2010). *The Turkish-Iranian gas relationship: Politically successful, commercially problematic*. Oxford Institute for Engergy Studies.

Koulaei, E. (2007). Joghrafiay-e Siasi va Eghtesadi Keshvarhay-e Hashiye-ye Darya-ye Khazar; Ta'amol va Hamkari ba Jame'e-ye Jahani (The geopolitics and geoeconomics of Littoral Countries of the Caspian Sea; Interaction and cooperation with global society). *Payam-e Darya (The Massage of Sea), 162*, 25.

Koulaei, E. (2008). Zheopolitic Enteghal Naft az Ghafghaz (The geopolitics of oil transmission from the Caucasus). *The Journal of Central Eurasia Studies, 1*(1), 77.

Koulaei, E. (2010). Naghsh-e Enteghal-e Enerzhi dar Ravabet-e Torkemanestan ba Russie (The role of energy transmission in the relationship between Turkmenistan and Russia). *The Journal of Central Eurasia Studies, 3*(7), 85.

Kouzegar Kaleji, V. (2007). Negahi be Bohran-e Ositia-ye Jonoubi va Hojoum-e Russie be Gorjestan (The taking of South Ossetia and Russia's invasion of Georgia). *The Journal of Political and Economical Information, 22,* 104–121.

Legvold, R. (2008). Russia's strategic vision and the role of energy. *NBR Analysis, 19*(2), 10–22.

Main, S. J. (2005). *The bear, the peacock, the eagle, the sturgeon and the black, black oil: Contemporary regional power politics in the Caspian Sea.* Swindon: Conflict Studies Research Centre.

Marketos, T. (2009). Eastern Caspian Sea energy geopolitics: A litmus test for the US-Russia-China struggle for the geostrategic control of Eurasia. *Caucasian Review of International Affairs, 3*(1), 2.

Mihalka, M. (2007). *Not much of a game: Security dynamics in Central Asia.* Paper presented at the China and Eurasia Forum Quarterly.

Mohammadi, F. (2009). Sofre-ye Khazar; Hamsayegan Mikhorand va Rezhim Nemigirand (The table of the Caspian Sea: Neighbouring Countries eat but don't diet). *Gostare-ye Enerzhi (Energy Range), 3*(27–8), 32.

Murphy, D. (2003). Asia's pipeline politics. *Far Eastern Economic Review, 6,* 13.

Nanay, J. (2001). Review of "Oil and geopolitics in the Caspian Sea region" by Michael P. Croissant. *The International History Review, 23*(2), 486–487.

Nosrati, H. R., & Etaat, J. (2009). Iran va Khotot-e Enteghal-e Enerzhie Khazar (Iran and the Caspian energy pipeline). *The Journal of Central Eurasia Studies, 2*(3), 4.

Ögütçü, M. (2003). Caspian energy and legal disputes: Prospects for settlement. *Oil, Gas & Energy Law Journal (OGEL), 1,* 2.

Oktava, O. Z. (2005). American politics towards the Caspian Sea and Baku-Tbilisi-Jiyhan. *Perception, 17,* 17–34.

Olcott, M. B. (2006). *International gas trade in Central Asia: Turkmenistan, Iran, Russia and Afghanistan, natural gas and geopolitics from 1970 to 2040: 375.* New York: Cambridge University Press.

Pabst, A. (2009). Central Eurasia in the emerging global balance of power. *American Foreign Policy Interest, 31,* 175–176.

Ramakant, D. (2006). China's Central Asia policy in recent times: China and Eurasia. *Forum Quarterly Central Asia-Caucasus Institute & Silk Road Studies Program, 4,* 147.

Sander, A. G. (2006). US-Russian energy security in the Caucasus: The prospects for conflict in the national interest. *The Journal of Russian and Asian Studies.* At: http://www.sras.org/us-russian_energy_security_in_the_caucasus. (November 25, 2017).

Saunders, P. J. (2008). *Russian energy and European security.* The Nixon Center.

Scaroni, P. (2006). Europeans must face the threat to energy supplies. *Financial Times, 18*(2006), 15.

Shadivand, G. (2013). Rouykard-e Etehadie-ye Ouropa dar Ghebal-e Naft-e Kham va Gaz-e Tabi'e Asiay-e Mian-e va Ghafghaz (The European Union towards crude oil and natural gas of Central Asia and Caucasus). *Journal of Exploration and Production, 98,* 40.

Shilibekova, A. (2008). Turkey-Kazakhstan relationship in the military sphere: Outcome and outlook. *European Journal of Economic and Political Studies, 1*(2), 57.

Shouri, M. (2007). Naft-o Gaz-e Khazar; Taghabol-e Mnafe' (Oil and gas of the Caspian Sea; The contrast of interests). *Asian Energy, 8.* At: http://www.isrjournals.com/fa/special-articles/1083-special-article-farsi53.html. Accessed on December 12, 2017.

Terterov, M., Pool, J. V., & Nagornyy, S. (2010). Russian geopolitical power in the black and caspian seas region: Implications for Turkey and the world. *Insight Turkey, 12*(3), 191–203.
Thomas, T. L., Shull, J. (1999). Russian national interests and the Caspian Sea. *Perceptions: Journal of International Affairs, 4*(4), 75–96.

Dayyan Janbaz (Iran) is a senior lecturer in International Relations at Payam-e Nour University in Babolsar (Hadishahr Unit), Iran. He holds a B.A. in political science from Mazandaran University, and an M.A. and Ph.D. in International Relations from Tehran University. His areas of interest include the Caspian Sea Legal Regime, Caucasus and Central Asian Studies, and Iranian Foreign Policy.

Mehrdad Fallah (Iran) is a fellow of the Faculty of Politics and Law at the University of Mazandaran, Iran. He holds a B.A. in political science, and an M.A. in International Relations from Mazandaran University. His interests include Iran's foreign policy and Middle East studies.

Chapter 6
The Separation of Bahrain from Iran

Masoumeh Rad Goudarzi and Sediqeh Nazarpour

Abstract The decline of Iran's Qajar dynasty, followed by a consequent politico-military crisis during the Pahlavi era, is associated with the firm establishment of British hegemony in the Persian Gulf with a view to guaranteeing India's security. At the same time, the discovery of oil and its importance to Britain's military and global economic empire doubled Britain's resolve to dominate the southern edge of the Persian Gulf. Hence, this paper investigates Bahrain's role in Britain's economic, political and military diplomacy in the region, along with the tactics Britannia employed to gradually separate Bahrain from Iran. The authors also disclose Iran's failed efforts to retain territorial integrity. Critical analysis suggests that by undermining traditional Iranian authority in Bahrain, Britain managed to protect India's western border while also gaining control over maritime traffic and the development and distribution of the region's oil resources.

Keywords Bahrain · Separation from Iran · British diplomacy · Persian Gulf
Oil resources

6.1 Introduction

For various geopolitical-strategic reasons, European competitors in the Middle East have always considered the Persian Gulf a commercial thoroughfare of vital importance. After the discovery of the region's oil, this importance grew logarithmically from purely 'transit access' to lucrative investment and global military concerns. Due to its historical rule over the Gulf of Persia's northern coasts and its dominion over the Channel Islands at the mouth, Iran has played an important role in Persian

M. Rad Goudarzi (✉)
Department of Political Science, Faculty of Literature and Humanities, University of Guilan, Rasht, Guilan, Iran
e-mail: fr.raad1@gmail.com

S. Nazarpour
University of Mazandaran, Pasdaran St, 741613534 Babolsar, Mazandaran, Iran
e-mail: sediqeh.nazarpour@gmail.com

M. Ebrahimi et al. (eds.), *The Dynamics of Iranian Borders*,
https://doi.org/10.1007/978-3-319-89836-0_6

Gulf security (Ramazani 1979: 6). Various regions of Iran have been lost over the last two hundred-plus years to newfound identities. However, none was as harshly borne as the loss of the Bahrain Archipelago, which had long been part of Iran's vast dominion. Bahrain has many ethnic and religious commonalities with Iran, such that the internal affairs of the emirate were and are greatly influenced by events in Iran (Daemi 1999: 1). The emirate's Shiite majority, ethno-Persian population, and rulers at the southern edge of the Persian Gulf, nearing Qatar and Saudi Arabia, comprise traditional ethno-Persian territories and peoples who, for centuries, paid homage to Iran's central government. However, after British colonialists imposed their presence during the nineteenth century, Bahrain gradually broke ties with Iran. Finally, with the political fall of the second Pahlavi dynasty, the islands separated from Iran by UN decree in 1971, thus, achieving a similitude of independence. Britain, via its East India Company, had always sought access to India and made every effort possible to append the Gulf region in favour of its 'Protectorates'. Attempts to 'liberate' southern emirates and convert them to independent states proliferated by means of signed protectorate agreements, each of which advanced British interests.

During its more than 150-year presence in the region, Britain always considered Bahrain to be independent from Iran, and due to the weaknesses of Iran's regional central governments, Britain finally managed to separate Iran's Fourteenth Province from its traditional liege. The main thrust of this study examines British goals as causes for the severed historical, political and economic links between Iran and Bahrain. Therefore, the paper attempts to study the role of Britain in both the Persian Gulf and the Indian Ocean, and examine Britain's *modus operandi* against Iran during the eighteenth, nineteenth and twentieth centuries. The main purpose of this study is to assess benefits gained by British exploitation, as well as expose Britain's subversive tools and policies.

6.2 The Strategic Location of Bahrain in the Persian Gulf

Bahrain ('A Million Palm Island') is an archipelago located in the southern Persian Gulf in a small bay called Shelva. Before Islam, the term 'Bahrain' referred to all southern lands in the Persian Gulf, from Oman to Basra. From the thirteenth century and during the Safavid era, southern Gulf hinterlands were politically mixed in with the Bahrain Islands. At the time, Bahrain contained the Avaal islands and hinterlands of al-Hasa (Kernoker 2011: 224; Partovi Moghaddam 2006: 81). During the Islamic middle ages, a city existed at the current site of Manama that was a thriving centre of trade (Ibn Battuta 1997). The Bahrain archipelago neighbours Qatar (on the east) and Saudi Arabia (south and west), with the Persian Gulf to the north, between the southern coastal city of Bushehr in Iran and the Arabian coast. It comprises an area of 698 km^2 with a population of about eighty-one per cent Muslim (seventy per cent Shi'a), nine per cent Christian and the remaining ten per cent being 'other' religions. The government is a constitutional monarchy with a King selected from the al-Khalifa family legacy (royal family) that took power in 1820 while announcing its alliance with Britain. In 1970 the country joined the United Nations and in 1971 it declared independence from Britain. Bahrain is a member of the League of Arab States and Cooperation Council for the Arab States of the Gulf. The US Navy's Fifth Fleet is based in Bahrain and, geopolitically, the country falls under the influence of Saudi Arabia's tightly knit security apparatus.

6.2.1 Bahrain in the New Era

In recent centuries, the Portuguese were the initial colonialists to achieve wealth east of the Cape of Good Hope. They entered the Indian Ocean and Persian Gulf regions, including Hormuz and Bahrain (in 1521), occupying them for almost a century. After the reign of the Safavid, Shah Abbas, political stability was re-established and Bahrain was relieved of its Portuguese occupiers in 1602, whereupon the central government's authority was set on shores and islands in the Persian Gulf (Kernoker 2011: 225). Thereafter, Bahrain remained under the full control of the Iranian government for the next 180 years (Valdani 1991: 311). The fall of the Safavids brought civil war and bloody confrontations, which distressed the political affairs of Iran with the formation of local regimes and various Arab tribal dominions on Iranian coasts. In 1782 or 1783, Sheikh Ahmed Bin Khalifa, from the tribe of Bani Utbah and the Khalifa Family that had migrated from Najd in central Arabia to Kuwait, attacked the territory and conquered it along with Hula Arabians who had occupied Bahrain for several years. The British colonial government fully supported the Khalifa Family's (al-Khalifa) rule in Bahrain (Valdani 1991: 311; Wilson 1987: 283). When Nader Shah came to power, his efforts suppressed local authorities and the situation changed in the Persian Gulf. He ordered the Fars ruler to attack Bahrain. The Dutch refused to help but Iran's forces prevailed and conquered the island. In the absence of the Sheikh of Bahrain, the Iranian commander sent the key of the Castle of Bahrain to Nader (Houshang Mahdavi 2009: 110). Afterwards, Agha Mohammad Khan Qajar founded the Qajar monarchy, strengthened its authority, and appointed Sheikh Nasr Khan, a local tribal chief, to rule Bahrain. In early 1797, a few months before his assassination, Agha Mohammad Khan ordered Sheikh Nasr Khan to provide security for the Persian Gulf and to punish the ruler of Muscat who had encroached on Persian territory. However, the sudden death of the Qajar Shah changed the plan, and the Sultan of Muscat remained ruler of the Persian Gulf (Zarrin Qalam 1958/1337: 57).

Iran's government directly influenced Bahrain's political situation. What distinguishes this era from others is the ubiquitous presence of Britain in the Persian Gulf and Bahrain (Valdani 1997: 300–301). In early 1819, Commodore Bruce, commander of the British fleet in the Persian Gulf, made a pact with Hossein Ali Farmanfarma, the Fars governor, which was to last "*until the time when the Iranian State cannot protect the Persian Gulf, it is the responsibility of the British government*." Thus, Iran waived one of its most important and inviolable rights in the Persian Gulf (Ghaem Maghami 1982: 16). Following the signing of the contract, a fleet of six war ships and three thousand sailors headed to the Persian Gulf from Mumbai. Sir William Grant Kamier was fleet commander. He conquered Ras al-Khaimah by force of artillery, after which he conquered the remaining southern Persian Gulf emirates. Like dominoes, they all fell without serious resistance. After this campaign, the British military commander negotiated with each emirate. On 6–7 January 1820, he signed treaties with eleven Arab sheikhs who pledged to end conflict and piracy with each other. The Sheikh of Bahrain also participated, which introduced the British Protectorate to the region. On 31 May 1861, the Bahrain archipelago—in which oil had recently been discovered—officially became a protectorate of the British government (Ghaem Maghami 1982: 16). During the 20th century, Britain consolidated its influence and interests in the Persian Gulf, including Bahrain. WWI brought a wider presence of British troops in the Gulf to strengthen Britain's position in the region. To further consolidate Britain power and influence, British officers residing in Bahrain forever justified their actions with excuses such as maintaining Gulf security and/or supporting Iran (Kernoker 2011: 226). This situation lasted until the reign of Reza Khan. During the Pahlavi era, Reza Shah invaded Khuzestan and put an end to Sheikh Khaz'al's rule, after which he referred problems between Iran and Bahrain to The League of Nations. However, by November 1927, the effort led nowhere. When Britain decided to leave the Persian Gulf in the late 1960s, it proposed syndicating Qatar, Bahrain and seven emirates along the coast (UAE), to face Iranian opposition.

6.2.2 Britain's Presence in Bahrain

The United Kingdom's relationship with Bahrain began with its presence in southern Iran and the Persian Gulf region during the patriarchal Mughal era with an exchange of ambassadors and correspondence. When the Mughal patriarchy collapsed, this relationship came to a complete standstill. During the Timurid era, relations were renewed by their having to face the Ottoman Turks, a common enemy. With the establishment of the Safavid state, the political, economic, and military relationships of both countries entered a new cycle of growth and rebirth. This period saw multi-faceted movements of Britain in the Persian Gulf and Gulf of Oman. By the Qajar and Pahlavi eras, Britain's influence had stabilised and became an exclusive privilege. Britain then effortlessly separated Bahrain from Iran without cost. Thus, British presence in Bahrain and in the Persian Gulf region can be divided into four periods: (1) influence, (2) expansion, (3) established hegemony, and (4) the acceptance of Britain's hegemony over the region by European powers.

6.2.2.1 The Era of Colonial Influence: The Expulsion of the Portuguese from the Persian Gulf and the Rise of British Power

When Europeans arrived during the sixteenth century, Bahrain had been paying tribute to the prince of Hormuz. The Portuguese took Hormuz and then Bahrain during the Safavid era, after which the Shah and nobility of Hormuz were gradually excluded from economic and political affairs. The main effect of Portuguese policy was the breaking of economic bonds between the island's strong families and their political influence. The complete Portuguese dominion of economic affairs caused major gaps in the community's traditional social composition. Gradually, the major clans vanished from seats of power, essentially because local capitalists grew indifferent to their fate (Vosoughi 2007/1386: 770). On the other hand, due to rigid pressures from Portuguese authorities, including heavy and irregular tariffs and taxation, merchant ship owners became reluctant to berth in Hormuz and chose Jask instead. Furthermore, the presence of Dutch and British powers in the Persian Gulf, in addition to the re-emergence of Iranian power under Shah Abbas I (1622), eventually led to Portuguese expulsion from the Persian Gulf (Monshi 1998: 1622–1623).

After the Portuguese navigator, Vasco Da Gama, discovered the passage to India from South Africa in the late fifteenth century while seeking trade with the Orient, European maritime countries such as Portugal, the Netherlands, France and Britain reached the Persian Gulf. The Dutch, French and British nations each established a company in India and looked to the Persian Gulf as a marine highway. Britain established its initial commercial ties with Shah Abbas I through the Shirley Brothers in 1600, and they acquired use of the port of Jask as a naval base to resupply British vessels. However, Shah Abbas ordered that no facility could be held at the port and that Iran would later claim ownership of the entire region (Moghtader 2007: 35).

The history of official British involvement in the Persian Gulf began in 1616, with the establishment of its commercial base at Jask. From then to the late eighteenth century, British activities were predominantly commercial. In 1616, Shah Abbas I granted a concession to the British East India Company by which they were permitted to trade within the country, allowing future company representatives to enter any Iranian port. Three years later, the company established its business centre in Jask (Curzon 1994: 625), chosen because of its proximity to the Strait of Hormuz and because it could be easily defended against Portuguese attack.

Reasons for the Shah's cooperation with the company include the following:

A. Iranian silk exports, previously taken through Aleppo, were diverted to the Persian Gulf to deprive the Ottoman Empire of customs revenue;
B. To extend effective Iranian rule over Persian Gulf islands and beaches, the Safavid Shah needed the collaboration of the British East India Company to thwart the Portuguese.

With the gradual exit of Spain and Portugal from colonial rivalry, a favourable context was prepared for the Dutch and particularly for Britain in the Persian Gulf. In January 1622 a military alliance between Iran and the British East India Company expelled the Portuguese from Hormuz. A joint attack on 23 April 1622 defeated the Portuguese and removed their flag after a hundred years of dominance. Afterwards, the Hormuz Customs House relocated to Bandar Abbas and British company officials transferred their business centre from Jask to Bandar Abbas. At the time, the British had also established another branch in the port of Basra. All contemporary trade with Europe to India ran through the Mediterranean, Syria, Iraq and the Persian

Gulf. The importance of the conquest of Hormuz during the reign of Shah Abbas is that it benchmarked the beginning of Britain's sweeping economic hegemony over the Persian Gulf. In addition, the replacement of Hormuz with Bandar Abbas as a central port in the Persian Gulf led to the relegation of Hormuz to a beach in the longer term (Vosoughi 2007/1386: 550). As a result, these potential readjustments actually minimised Iran's continuing sovereignty over parts of the southern coast of the Persian Gulf and Sea of Oman.

6.2.2.2 The Era of Colonial Expansion: Increasing Power for Local Tribes and the Undermining of Iran in the Persian Gulf

The eighteenth century saw this era begin in earnest. With the fall of the Safavids, the lack of a strong central government in Iran allowed Persian Gulf coasts and ports to be overrun by clans and tribes. Violations by the Sultan of Muscat (Ibn-Saif), which began with the Safavid dynasty, grew to include Bandar Abbas through to Bahrain (Wilson 1987: 201). The fall of Isfahan in 1722–23 cast the shadow of death over the body of Iran's maritime economy. On the one hand, it provided fertile soil for the growth and development of local Persian Gulf tribes; on the other, it facilitated the rapid demise of Iranian power in the Persian Gulf. During this period, Iran's navy weakened, only to be replaced by ships from the Imam of Oman and those from authorities in Bahrain, Basra, and Sharjah. Hence, after defeating the Afghans and his consolidation of power, Nader Shah Afshar came to realise the importance of the fleet. The Port of Bushehr then became Iran's naval base in the Persian Gulf. Nader Shah sought to reform and consolidate his naval power in the Persian Gulf, and even managed to take Muscat and Oman, and once again raise the Iranian flag over Oman (Mojtahed-Zadeh 2006: 24–25). However, British and Dutch navies disapproved. Thus, Nader Shah Afshar did not achieve further advances. Generally, despite decent political and economic relations with Iran during the period of Afshars, Britain's East India Company opposed the formation of an independently powerful Iranian fleet in the south. Its intent was to isolate Iranian capability in the region and increase its own profitable monopoly (Chamankar 2000: 2). The importance of the Persian Gulf, and particularly Bahrain, especially in terms of trade and Britain's protection of its Indian interests, trumped Bahrain's cooperation with other major powers, making for a tense continuum in the region throughout the eighteenth century (Elahi 1991: 66). To maintain this tension, divisions between Bahrain's tribes prevented unification and any such regional initiative.

British influence expanded with the advent of the Zand dynasty. Undoubtedly, (i) the Afghan invasion of Iran, coupled with (ii) the fall of the Safavid dynasty and (iii) emerging Ghasemi dynasty in the Persian Gulf, plus (iv) Karim Khan's ongoing development of foreign trade, led to increased British influence in the region. Signing contracts with senior officials of the British East India Company, Karim Khan allowed the company to build a large business centre in Bushehr. The British Resident and agents were then transferred to Bushehr (Wilson 1987: 233), and remained there until 1946, when Iranian officials 'urged' their relocation to Bahrain. This same era also ushered the decline of the Dutch in the region, thus, removing major obstacles from the advance of Britain's dominance when entire French and Dutch colonies were 'reassigned' to the British East India Company, making Britain the absolute power without rival in India. This circumstance further elevated the importance of Bahrain and the Persian Gulf in terms of economy and transport. The approximate travel time from Aleppo to Basra and the Persian Gulf, and from thence to India, took five months by land and ship, compared to rounding the Cape of Good Hope, which took about eleven months. Therefore, rerouting via the Persian Gulf reduced the journey by six months, which was more economically viable.

6.2.2.3 The Qajar Era: Continued Rise of Local Tribes and the Establishment of British Hegemony Over the Persian Gulf

During the early nineteenth century, the Qajar dynasty, for internal and external reasons, was unable to intervene in Persian Gulf affairs directly. After solving problems in the Caucasus, and following Agha Mohammad Khan's death, riots, civil strife and Iran's war with Russia created a rather complex situation that immensely influenced the Persian Gulf (Eghtedari 1991: 142). In addition, Napoleon's ascent and attacks in the East, along with his close relationship with the Imam of Muscat, were other factors affecting Britain's influence in the region. In fact, Persian Gulf competition among the great players of the time (Tsarist Russia, France and Britain) was especially keen. Russia's Peter the Great wanted warm waters, Britain lusted for economic and colonial gains as France reached towards India in a desperate act of compensation for prior failures. All flocked to Bahrain and the Persian Gulf. Their collective presence fixed all eyes on Iran, which then led Britain to study Iran with greater fervour (Vosoughi 2007/1386: 142). Moreover, Al Saud's ascent encroached on Persian Gulf shorelines, which led to the independent strengthening of several local tribes. All of this heightened Britain's sensitivity to the Persian Gulf, causing a rash of activities that brought new challenges to Iran (Mojtahed-Zadeh 2006). Tribal sympathies in the southern Persian Gulf leaned towards British Protectorate coin and force (Sadid Al Saltane 1991: 99–100), which, from the beginning of the nineteenth century, served to stabilise British influence in Muscat and Oman. Contemporary concerns were of sufficient scope for the British government policy to embrace collaboration with tribal chiefs, allowing the purchase of influences that promoted the fixation of Britain's dominion in the Gulf, especially given the weakness of Iranian naval installations (Houshang Mahdavi 2009: 302–303). Iran's status gradually and completely eroded amongst regional tribal councils. The dynamic continued during the Qajar dynasty.

From the beginning of the nineteenth century, Britain used the Persian Gulf, by far the safest sea route, to protect its interests in India. Hence, Bahrain held great importance as a base, therefore the British took extraordinary measures to negotiate contracts with carefully chosen Sheikhs who followed protocols that protected British interests in the region. Britain's first contract with local sheikhs in the Persian Gulf was signed in 1798. This was the Treaty of Alliance between the

British East India Company and the Sultan of Muscat on behalf of Oman. Britain later signed another key contract in 1820 with several sheikhs of al-Khalifa. This agreement permitted Britain—'during a period of confusion and insecurity'—to continue its military presence in Ras al-Khaimah as the gateway to the Persian Gulf. Based on this rather vague pretext, it deployed warships around the island as its political representatives stayed in the emirates, especially Bahrain. Hence, Britain shared in the region's return to stability during any insecurity or civil disorder, thereby justifying its presence. Bahrain was actually forced to join the 'substantial agreement' in 1820 (Elahi 1991: 68). Sheikh Salman Bin Ahmed, Sheikh of Bahrain, seized the opportunity to strengthen his power and authority under British protection. The British flag thus achieved its strongest position in Bahrain to date (Ghaem Maghami 1982: 15). However, the provocation countered the Anglo-Persian treaty signed a few years previously. Nevertheless, Britain subsequently consolidated its dominion further with its Treaty of Perpetual Peace with Arab sheiks in 1853. Under this contract, the British were permitted to interfere in any untoward incident along Persian Gulf coasts. The emirates formally accepted Britain's support and the latter signed additional 'protectorate agreements' with Muscat and Kuwait to consolidate its control of the region. Under the pretext of combating German or Ottoman threats, it occupied coastal regions and islands in the Persian Gulf, including the three islands of Abu Musa and the Greater and Lesser Tunbs, thus, limiting Iran's power in the Persian Gulf. The implication is that, to forestall Arab unity, the late nineteenth- and early twentieth-century British policy was deliberately designed to develop numerous smaller emirates in the Persian Gulf while subjugating them. Finally, Britain took control of Bahrain in 1861 and, from 1880, began its invasions of the four Siri, the Greater and Lesser Tunbs, and Abu Musa Islands. From this time onwards, Britain significantly changed its policy from economic to political while continuing the pretext of defending its Indian and Gulf interests. Hence, the British government and its East India Company gradually began to interfere in Arabian political affairs (Eghtedari 1991: 42; Kia 2006: 2–12).

6.2.2.4 The Pahlavi Era: Acceptance of Britain's Hegemony Over the Region by European Powers

Multiple factors increased Bahrain's importance during this era, the most important of which was the discovery of oil in 1908 (Ebrahimi 2016), which allowed Britain to change the power source for its commercial and military fleets from coal to oil. The next most influential factor was the 1917 Russian Revolution, which placed Russia at ideological odds with Britain. Bolshevik policies and propaganda promoted a global uprising of colonised nations that directly challenged British imperial interests. Another important element was America's entrance into the ring of international competition, especially after WWII, when it rivalled UK interests in the Persian Gulf. During the 1920s, competition waxed between Britain and France, on one side, and Germany and Italy on the other. Americans, who predicted the European war, exploited advantages offered by the economic collapse of warring European countries, and the United States became the most powerful country in the world.

By this time, Bahrain had gained far more importance in UK eyes as part of its regional strategy. Beginning in Qajar, it undertook several initiatives to weaken Iranian influence in Bahrain, all of which were finally completed during the Pahlavi era. Among these were the formation of an Arabian government in the Persian Gulf, the bestowal of a Medal of Honour on the Sheikh of Dubai, and the resolution of numerous inter-Arab disputes. British contracts with the emirates granted Britain the office of 'Protector', while its use of military and naval forces, in addition to shrewd and effective diplomacy, hindered all Iranian efforts to protect its own interests in the Persian Gulf (Khosravi and Alimardani 2011: 170). British pressure tightened in Bahrain and its interventions were such that it even prevented the teaching of the Persian language in Bahrain's schools. Funds also flowed from the British government to establish regional military, defence and observatory posts to monitor all traffic through the Gulf. On 13 September 1924, correspondence between the Iranian Foreign Ministry and Ministry of War mentioned the arrival of British troops in Bahrain (Khosravi and Alimardani 2011: 170). To consolidate its political and economic positions further, Britain brought warships into the Gulf and troops to Bahrain in complete violation of Iran's rights in Bahrain. These activities advanced so well that Britain's agents publicly spoke about Bahrain's independence.

In response, Reza Shah resurrected Iran's claim to Bahrain and referred the problem to The League of Nations in November 1927, but the measure went nowhere. Iraq then became a formal British protectorate on 9 May 1928. The ongoing border disputes between Iran and Iraq, especially over the Shatt al-Arab River, were major factors that effectively supported the measure due to a lack of effective Iranian sovereignty in Bahrain. After WWII, Iran again raised claims to sovereignty in Bahrain and referred it to the UN. On 11 November 1952, Iran even introduced Bahrain as its fourteenth province. However, the withdrawal of British forces from the Gulf, and the theory of dominant regional power as discussed by Arab States at the time, tempted Iran to ignore its sovereignty over Bahrain in lieu of control over the Gulf region and sovereignty over three Strait of Hormuz islands, the Greater and Lesser Tunb and Abu Musa. On the other hand, the British government, via its dependent Iranian agents, began discussions on the unimportance of Bahrain among Iran's courtiers and politicians. They posited the depletions of Bahrain's oil reserves and a decline in the global market for pearls. The Shah, who wished to be gendarme in the Middle East, then sought negotiations with Britain. However, in the late 1960s, the British decided to leave the Persian Gulf to its American operatives and presented a syndicate comprising Qatar, Bahrain and seven Persian Gulf emirates (UAE), concerning which Iran adamantly opposed the inclusion of Bahrain (Arlowe 1962: 22).

In this regard, the Saudi King stated that Arabs should replace Britain and he welcomed Isa Bin Salman al-Khalifa to Riyadh as Bahrain's new independent ruler. This immediately caused a dispute between Saudi Arabia and Iran, upon which Mohammad Reza Shah cancelled a meeting with the Saudi King (Mojtahed-Zadeh 2001: 158). Finally, in 1968 the Shah announced that Iran's sovereignty over Bahrain was subject to the will of the island's people, and sent a letter to this effect to the UN Secretary-General calling for a referendum. After a two-week survey, the UN special representative reported '*the majority of the people of Bahrain want independence.*' Accordingly, the UN Security Council's Resolution 278 (11 May 1970) recognised the independence of Bahrain. Hence, UK efforts finally succeeded in the complete legal separation of Bahrain from Iran.

6.2.3 Separation of Bahrain and British Interests

Due to the actions cited and ongoing British efforts to secure Bahrain's independence, we must ask, 'Why was it so important?' There are several reasons.

Economic: Bahrain is an ancient hub of global trade with India, China and South East Asia through to the Ottoman Empire and Europe. Europeans, attracted by the region's great wealth, had long been established there until Britain's navy entered the Gulf to defeat colonial states like Portugal and Holland. After weakening indigenous regional governments, Britain sedulously extended its control. Pending a strong central government that could provide security, Britain did not intervene in Bahrain's politics except for commercial activity. However, following the Russo-Persian War and the subsequent ascent of southern Gulf piracy, Britain began to secure its maritime routes to India and gradually established its regional naval strength, especially in the absence of any Iranian naval power. At this stage, British goals were to protect its commercial interests and reduce the influence of other foreign powers in the Persian Gulf.

Long before Atubah, and traceable to the Safavid dynasty, beaches and islands in the region were of political interest to all colonials (Ghaem Maghami 1982: 2). Before Nadir Shah Afshar, Britain only exploited commercial programmes. However, from 1750 to 1763, Dutch ambitions for Bahrain caused Britain to enter indigenous politics and rub shoulders with local sheikhs. Britain moved its East India Company's HQ (EIC) from Bandar Abbas to Basra, which began a new colonial enterprise era in the Gulf. During the second half of the eighteenth century, British agents then initiated relations with the ruler of Bushehr, Sheikh Nasr Khan (Bahrain was under the administration of Bushehr), to strengthen their positions in Bushehr and Bahrain (Amin 1981: 165–169; Deldam 1984: 100). In 1793, Britain also signed a cooperation agreement with the Sheikhs of Bani Utbah in Kuwait, which can be considered the link that stimulated al-Khalifa to take control of Bahrain. WWII brought great losses, with urban destruction, widespread unemployment, inflation, industrial chaos and Britain's inability to repay heavy debts to the US. The Persian Gulf's thriving underground markets, both north and south of the Strait of Hormuz, had the potential to relieve part of Britain's financial haemorrhage.

India: Initially, Britain's hegemony in the Persian Gulf and Bahrain protected its Indian interests from rival European powers, as well as from Ottomans and Persia, who might otherwise have used the Persian Gulf to invade India. Hence, Britain paid enormous attention to the Gulf and kept regional countries enfeebled by the politics of dissension. Preserving India—the beating heart of the British colonial empire—was intimately dependent on strengthening Britain's defensive positions in the Persian Gulf. India was at the very heart of British foreign policy, and any issues that arose in Europe pertaining to the economy, military and otherwise, orbited its Indian preserve. Lord Curzon considered India a non-negotiable matter concerning British rule in the East, and Churchill regarded India as the shining jewel in the crown of Britain's monarchy, such that its loss would cause the nation's collapse (Modak 1966: 121; Nehru 1971: 730–731).

Hence, because the Portuguese had a strong presence in southern Iran and due to Iran's naval weakness, Britain signed the Minab agreement with Shah Abbas to secure a strong foothold from which it could expel the Portuguese troops that had greatly distressed the local populace. According to the agreement, spoils of war were divided equally between Iran and the United Kingdom while imported British goods were exempted from customs duty. This contract marked the beginning of a far more significant British presence in southern Iran, along with influence in other regions such as Bahrain. Accordingly, Britain's preservation of the 'Indian Jewel' was its foreign policy basis. Any movement in the Middle East, especially by Iran, was therefore considered 'sensitive' by London (Elahi 1991: 44).

Bahrain's oil: Bahrain's oil was reason enough for Britain to ensure its separation from Iran. The Persian Gulf gained additional value with the discovery of oil in 1908, including in Iran and other countries. This allowed Britain to convert its fleets to oil-driven vessels, reduce operational costs and increase manoeuvrability. Hence, the importance of Iranian oil increased because otherwise London had to purchase oil from other countries, including America and Mexico. Therefore, complete domination of Iran's oil as a fuel source for its warships became a British priority. By 1919, with the development of larger engines and aircraft, oil had become the major source of energy. By 1932, Bahrain produced its first oil and exported forty thousand tons by 1934. Later exports increased to more than one million tons annually. By 1938, the Persian Gulf and the Middle East supplied five per cent of the world's oil consumption, and 22 per cent of UK energy needs. This alone prompted Britain to take exclusive control of Bahrain's oil until the independence of Bahrain in the 1960s. Therefore, oil's importance to the international economy and geopolitical balance was one of the most important reasons for the separation of Bahrain from Iran. British priorities in Bahrain and the Persian Gulf can be summarised as follows:

1. Protection of the western borders of India and Karachi from external aggression;
2. Protection of maritime traffic in the Persian Gulf;
3. Protection of oil fields, pipelines, and oil refineries in Bahrain and Abadan;
4. Protection of its regional trade and commercial interests.

6.2.4 Methods Used by Britain to Dominate Bahrain and Separate It from Iran

The most obvious colonial tool was that of making contracts with Bahrain's ruling sheikhs while disregarding Iran's central government. The most influential of these was the Perpetual Truce of Peace and Friendship, signed in 1820 between Britain and eleven independent sheikhs and tribal chiefs, as well as Bahrain's Sheikh, which marked the official beginning of British interference in Bahrain. The contract directly countered the provisions of the 1814 agreement between Britain and Iran. According to that detailed treaty, all political relations between Iran and Britain in Afghanistan, the Persian Gulf and matters regarding political fugitives were specifically cited to prevent Britain's unlawful interference or its establishment of 'private relations' with local tribes and rulers. Article 2 of the treaty states:

> When there is conflict between tribal chiefs and border rulers, Britain has no right to interfere, and if rulers offer a part of Iranian land to Britain instead of help, the British government must refuse and not interfere in Iran's internal affairs. (Eqbal Ashtiyani 2005: 134).

Despite these promises, the keynote feature of the 1820 contract was that the Sheikh of Bahrain signed the document without the permission of the Iranian court. At the time, Bahrain was part of the Fars Province, and any such action required the confirmation of the Fars ruler. Hence, this agreement was the first step taken to subvert Bahrain's political relations with Iran (Momeni and Zamannejad 2011: 213), and appears to have had the most far-reaching consequences. In March 1820, Henry Willock, British Foreign Minister, challenged Iran after the disappointment of Iran's refusal to permit British bases on its shores in the Persian Gulf. He actually threatened to occupy Bahrain (Farahmand 2007: 149; Movahhed 2001: 45–46).

Britain's *de-Iranization* policy should be considered an attempt to politically dominate the East, specifically in terms of a business strategy, whereby Britain actually deigned to keep Iran isolated and weak. Hence, strengthening southern Persian Gulf sheikhdoms and manipulating the removal or installation of Arab sheikhs disregarded Iran's authority and local dominion. We trace Britain's open intervention in Bahrain to 1814, when the East India Company, on behalf of Britain, reached an agreement with Sheikh Abdullah Bin Ahmed to aid him against Iran, whereupon the Sheikh implemented all British demands in exchange. This pact set the tone for all future agreements. The final blow came with a treaty between Colonel Bruce and Sheikh Isa, Bahrain's ruler, on 22 December 1880:

> I, Isa Bin Khalifa, Sheikh of Bahrain, hereby, on behalf of my own and my people, pledge to the British government not to negotiate or enter into any commitment with governments other than Britain; and refuse to allow the establishment of political or consular representation to any other state except for Britain; or allow military alliance matters within my territory except with the permission of the British government. (Adamiyat 1954: 102)

In this regard, by the late nineteenth century, British dominion was complete. Bahrain even proposed reasons for rejecting Iranian sovereignty as follows:

1. Bahrain is geographically distant from Iran, separated by sea, and closer to the coast of Oman.
2. Sheikhs in the region, especially the Sheikh of Bahrain, are independent rulers.
3. Residents of the territory speak Arabic and are not a part of Iranian race.
4. British intervention—vis-à-vis commitments and agreements with local sheikhs and rulers—in the absence of contracts with the Iranian central government proved that local leaders were not dependent on Iran. All contracts were based on the assumption that any claim of sovereignty by monarchy (Iran or Ottoman) was impossible, as established by the Government of His Majesty, the King of England.
5. In 1822, the time of Fath-Ali Shah, an agreement was signed between Captain William Bruce, the commissioner for the British government in the Persian Gulf, and the prince of the Fars, Hossein Ali Mirza Farmanfarma, in which the sovereignty of Iran over Bahrain had been recognised. According to the British foreign secretary, since the contract was signed without the requisite permission, the same was refused immediately. Therefore, the British government did not approve it either, and consequently rejected the notion of Iranian sovereignty over Bahrain.
6. Britain announced that during part of the seventeenth century and the last few years of the eighteenth century, Bahrain had been occupied by troops from Iran and Persian Gulf rulers. However, it seems that in or about 1783, Bahrain's domination by Arab tribes demonstrated Iran's loss of the islands as possessions. Since then, these islands had never been under effective Iranian supervision. Hence, Iran had not applied its right of ownership since 1783, indicating Bahrain was no longer its possession.

Therefore, Britain's policy of *de-Iranization* and these justifications rejected Iranian sovereignty over Bahrain in favour of British interests. The background of regional disputes between Iran, Turkey and Oman on one hand, and Arab sheikhdoms on the other, also served to weaken Iran's position.

Promoting enmity between leaders and elders of the al-Khalifa was yet another efficient tool of British interference and hegemony in Bahrain. Britain assiduously studied relationships between the local sheikhs and Fars governors and managed to appoint its own favourites or change circumstances that promoted their traditional 'divide and rule' politics. For example, Sheikh Abdullah, a strong man, ruled nearly twenty-two years and prevented British interventions in Bahrain. Britain managed to provoke local leaders to attack Bahrain by sea, forcing Sheikh Abdullah to flee (Neshat 1971 : 458–459; Sadid Al Saltane 1991: 460).

Moreover, correspondence between Sheikh Mohammed Bin Khalifa and the Iranian court expressed his loyalty to Iran's government. The British Government felt endangered and changed the situation to its advantage. Accordingly, they provoked the Sheikh's brother, Ali, to act against him (Neshat 1971; Saltane 1991), which led to a quarrel during which Ali was wounded. Asadollah Mirza, the ruler of Bushehr, in a letter sent to the Foreign Ministry, quoted the representative of the Sheikh of Bahrain, saying: "Since we raised the flag of Iran, and we know Bahrain as part of

Iran, ambassadors of the British government have laid a foundation of bad conduct and abuse. We are not comfortable with them, and we want to be safe in the shadow of the government of Iran" (Farahmand 2007: 150).

To prevent a unified front from rising against them, British efforts and policy deliberately and slowly created several smaller emirates in the Gulf (Elahi 1991: 72). They effectively assigned a rule of local puppets who pledged loyalty to British sovereignty. An example is Sheikh Hamad, son of Sheikh Isa, who wrote in 1921, "According to the order of Great Britain's government, from now on, I take the responsibility of the country" (Tolouee 1987: 122). From this time onwards, the bond between the Khalifa Family and Britain became so strong that the government of Iran was no longer able to assert its sovereignty over part of its traditional territory, the archipelago of Bahrain.

6.2.5 Iran and the Separation of Bahrain

Britain was determined that Bahrain would become independent, even though Iran called the little nation its fourteenth province and had claimed sovereignty for years. Nevertheless, as British influence decreased in the face of American ascendance in the Gulf, it did not want Iran, with its vast coastline, to influence the Gulf's opposite shores. Meanwhile, the Gulf sheikhs tried to prevent the exit of British troops and held numerous meetings to this end.

Britain had long consolidated its position of leadership in the Persian Gulf and Oman Sea and continually blocked all strategic Iranian hegemony in the region by any means possible (Chamankar 2000: 5). In 1861, the Bahrain archipelago became a British protectorate and in 1880 they began their 'invasion' of the four Siri, Abu Musa, and Greater and Lesser Tunb islands. The Governor-General of India, Lord Curzon from 1858 to 1947, reported the following to the Indian branch of the British Foreign Service and Commonwealth Office (21 Dec.1899):

> Great Britain's strategic interests in Iran stem from the close and cordial relationships of the country with India. From long ago, Indian borders have been limited to their current borders. However, Iran is vital to England vis-à-vis its influence over the seas south of Iran and the Sea of India, which pose possible risks for England. (Farid 1978: 12)

In addition, another report from George Hamilton, British Foreign Secretary, emphasised:

> We have many business interests in the Persian Gulf. We have contracts with the heads of the Persian Gulf, and they know us as a sponsor. The government of His Majesty the King of England warned the government of Iran, that ignoring the legitimate interests of Great Britain and India in ports in the Persian Gulf and southern parts of Iran will result in dire consequences. (Salimi 1922: 16)

British policy towards Iran and the Persian Gulf was also highlighted by the following telegram from Lord Lansdowne to Hardinge, British's ambassador in Tehran (06 Jan. 1902):

> We cannot imagine with any logic, that Britain closes its eye about its position, which has been earned through several years of efforts, or remain silent about efforts of other states to analyse political priorities in southern Iran. London has required all suitable measures to protect the interests of Britain, and measures through sea defences face no obstacle. His Majesty's Government, to protect its interests, which are at stake, if it is necessary, has the right to neglect the integrity and independence of Iran (Elahi 1991: 41)

WWI and the presence of British troops in the Persian Gulf presented an opportunity to consolidate Britain's position in the region. To strengthen power and develop influence, Britain used several excuses and pretexts to justify and maintain security in the Persian Gulf. British officers like Charles Belgeru took residence in Bahrain and others introduced themselves as assistants and supporters of the Iranian government. During the reign of Reza Shah Pahlavi, disputes between Iran and Britain over Bahrain concerned the following major themes:

1. *Bahrain's Oil*. In 1930, Bahrain's oil concession was given to a US company, Standard Oil of California. Iran considered Bahrain part of its territory and claimed its share of the oil by referring the issue to the international community (Hedayat 2006: 446).
2. *Passports*. In 1922, Iran informed all governors of Persian Gulf ports that the people of Bahrain held Iranian nationality and did not have permission to leave the country without an Iranian passport. Britain countered (Zargar 1993: 247), but the issue was stipulated once again in 1928 (Hedayat 2006: 328).
3. *Arms smuggling from Bahrain*. On 8 June 1930, Teymourtash, the court minister of Iran, sent a letter to Clive, the British ambassador:

   > When arms dealers can escape from officers of Iran, which has no navy, it is understandable. However, how can they pass through areas where British ships are deployed; ships that have the task of preventing arms smuggling?

4. *Bahrain sheikhs' actions against Iran*. Hedayat wrote:

   > From 1923–31, we received news that the government removed residents of Iran, destroyed their property documentation, closed Iranian schools, took tax bills out of the hands of sheikhs, and banned the Pahlavi hat (Hedayat 2006: 447).

5. *Stamp type envelopes*. The government of Iran ordered that Bahrainian envelopes must have an Iranian postal stamp, which met great opposition in British quarters.

Iranian protests were feeble and no real action was applied to restore Iran's sovereignty over Bahrain. With this in mind, an agreement was signed between Britain and Abdulaziz Al Saudi (1928), under which the King of the Hejaz was obligated to establish peaceful and friendly relations in Kuwait, Bahrain and Oman, and any other coastal region holding a special relationship with Britain. Signing this contract violated the territorial integrity of Iran. In a memorandum to the British government and subsequently to The League of Nations, Iran formally complained. In notes exchanged between the Government of Iran and the United Kingdom, the language and literature of Britain indicated an underestimation of the Iranian side by British agencies. Another exchange, between Abdolhossein Teymourtash, Minister

Plenipotentiary of Reza Shah, and Sir Robert Clive, British Minister plenipotentiary of Britain in Tehran, took place. This next round of negotiations concerned an agreement intended to solve all political, financial, commercial and land disputes for both countries, with Bahrain as the most important focus. During negotiations, Iran gave indications of its willingness to withdraw with flexibility pertaining to Bahrain. The Court Minister of Reza Shah preferred negotiations to be referred to the international community. Britain wanted Iran's unconditional release of any claims to Bahrain, Abu Musa, and the Greater and Lesser Tunb Islands. Therefore, negotiations ceased (Kernoker 2011: 226–227).

In 1930, the Government of the United Kingdom prevented Iranians entering Bahrain with a Bahraini passport. Iran's response, through formal correspondence and objections, led nowhere. In the same year, Major Frank Holmes, representative of the Anglo-Persian Oil Company, gained an oil exploration concession in Bahrain and immediately began drilling operations. Protests and complaints from Iran remained ineffective in The League of Nations because of US support for British policies. At the same time, the British began actively removing Iranian cultural symbols from Bahrain. Disaffected Iranian residents in Bahrain, unwilling to accept foreign domination, desperately pleaded with the Iranian government to prevent violations of their human rights and the destruction of personal property. Iranian rulers responded by recommending patience, dignity and tolerance, even as formal protests to Britain and The League of Nations went unheeded. Events in September 1941 and Iran's political climate set the problem of Bahrain aside for a few years. However, in 1948 Iran disavowed all arrangements made with Bahrain by other governments. However, after August that same year, with Iran's alliance with Britain as a member of the Baghdad Pact, Bahrain was forgotten. Any objective work was superficial, and only the Press occasionally mentioned Bahrain and Iranian rights (Mojtahed-Zadeh 2006: 195–196).

We see therefore, that before the 1953 coup, Iran sought to restore its governance and ownership of the Bahrain Island group a total of eleven times. Each time, British opposition managed to neutralise its attempts (Zargar 1993: 246). By 1956, a wave of popular opposition to foreign domination, along with claims by the people of Bahrain for their indigenous rights, led to protests and conflicts. Some were killed and leaders were arrested and deported (Asadi Azghandi 2002: 179). Following these events, the Iranian government passed an act in November 1957, which recognised Bahrain as "… an integral part and the fourteenth province of Iran forever." Britain's Government protested by claiming Bahrain was '*an independent Arab state*', as confirmed by the British House of Commons (Yarshater 2004: 509). Following the British confirmation of its position, Mohammad Reza Shah, under strong British influence, announced in *The Guardian* (August 1966) that "Bahrain runs out of pearl reserves in its coasts, and is not important any more." Again, during a visit to New Delhi (January 1969), he announced that Iran's monarchy did not want to seize Bahrain by 'force' and would leave its sovereignty to the free will of a resident referendum under UN supervision. This 'new status' went solidly against a background of continued protest over the separation of Bahrain from Iran. Bahrainians had even protested the attendance of Bahrain in the Federation of Gulf States (18 July 1969). The so-called Referendum was actually a '*dignified escape*' that denied Iranian sovereignty, being the fruit of an agreement between Mohammad Reza Shah, Britain and America. Its provisions were announced immediately to U Thant, Secretary General of the United Nations (March 1970). The Iranian government formally called on the '... good offices of the Secretary General to ask the opinions of the people of Bahrain.' Britain agreed. Subsequently, preliminary steps for Bahrain's separation from Iran were provided by an Iranian diplomatic group headed by Amir Khosrow Afshar (Iran's ambassador in London), who held talks with William Luce, the political representative of the United Kingdom in Bahrain.

According to secret compromises and conditions at the time, the King of Bahrain chose a political solution. The referendum was conducted but without public participation. Its conductors (official views of a group of business people dependent on the presence of Great Britain) stated that the majority of the people wanted independence. The UN Security Council passed Resolution No. 278 (11 May 1970), approving Bahrain's independence, and the country joined the United Nations. Praise and much appreciation were offered to Mohammad Reza for his toleration and acceptance of democratic principles in Bahrain. Members of the new Bahrainian Congress and Parliament, the majority of whom did not truly represent the people, were appointed much like agents and government officials with unconditional obedience to the will of the King. They all then passed legislation for Bahrain's separation from the territory in an extraordinary session on 24 May 1970, with 199 votes in favour and 4 votes against. Iran was the first country to recognise Bahrain's independence formally (Amin 2005: 65; Houshang Mahdavi 2009: 2). This is how Bahrain left Iran.

6.3 Conclusion

Tensions between Iran and Britain during the nineteenth century were mostly over the sovereignty of Bahrain. What we deduce from historical sources is that Iran's sovereignty over Bahrain was often accompanied by military force. However, the remoteness of Bahrain from Iran's central government, despite taxes paid to rulers in southern Iran and the racial composition of the region, confronted Iran's claim of sovereignty with difficulties. British colonialism in southern Iran and the Persian Gulf, along with competitive European cousins such as Portugal, Spain, the Netherlands and France, can be traced to the Safavid period. Nevertheless, British colonial intervention in Bahrain began with the pretext of (i) eliminating piracy and the slave trade, soon followed by (ii) arms smuggling to Iran, Afghanistan and tribes in Oman, as well as (iii) 'protection of British interests in the Persian Gulf and India.' After the treaty between Britain and Arab rulers of Bahrain, a new chapter began in these relations leading to Bahrain's status as a British protectorate. With the discovery of its oil reserves, Bahrain's importance intensified. British presence and influence in Bahrain grew proportionally with astute manoeuvres that purposefully isolated the region from Iranian sway. Due to military and economic limitations, not to mention the lack of potent diplomacy, Iran failed to retain sovereignty over Bahrain. At a meeting with the Security Council, Iran's representative had no choice other than the diplomatic gesture of recognizing Bahrain's independence. He also stated that his government looked forward to the welfare of the people of Bahrain while hinting that the basic human rights of Bahrainians of Iranian origin should be respected.

Iran's acceptance of the 'independence referendum' exchanged historical claims for a newly legal, pseudo-moral legitimacy. The opinion poll was but a pretext that justified globalist protocols. According to international convention, whenever at peace, a government can be asked to overlook its right of sovereignty over a part of its territory because part of the population does not speak the national language, or because of separation from the region by sea. Discussions on geographic and ethnic issues are pertinent when a government requests accession of a new territory. Aside from Iran's rights to the Bahrain Archipelago, racial and economic ties ranked higher than any relationships with other independent countries. However, after half a century of conflict with Britain, and following the secret compromise between Mohammad Reza Shah and Western powers, Iran formally withdrew from its position as Bahrain's gendarme and protector of its interests in the Western and Southern Persian Gulf.

References

Adamiyat, F. (1954). *Bahrain Islands: A legal and diplomatic study of the British Iranian controversy*. New York: F. A. Praeger.

Amin, S. H. (1981). *International and legal problems of the Persian Gulf*. London: Middle East and North African Press.

Amin, S. H. (2005). Sargozasht, Seresht va Sarnevesht-e Hameporsi (Trends, characteristics and fate of election). *Hafez Monthly, 17*, 58–68.

Arlowe, J. (1962). *Persian Gulf in the twentieth century*. London: Cresset Press.

Asadi Azghandi, B. (2002). khalije fars va Masaele an (Persian Gulf and its Problems). Tehran: Samt Publication.

Chamankar, M. J. (2000). Estemer-e Britaniya dar Mantaghe-ye Khalij-e Fars va Darya-ye Oman az Sheklgiri ta Payan-e Qajariyye (British Colonialism in the Persian Gulf and Oman Sea from the early stages to the end of the Qajar Era). At: http://bcir.pchi.ir/show.php?page=contents&id=9471. Accessed November 25, 2017.

Curzon, G. (1994). *Iran va Ghaziyye-ye Iran (Persia and the Persian Question)* (G. A. Vahid, Trans.). Tehran: Elmi.

Daemi, M. H. (1999). *Esteghlal-e Bahrain va Esterdad-e Jazayer-e Segane-ye Iran-i dar Khalij-e Fars (Bahrain independence and the restoration of the Persian Gulf Triple Islands)* (Unpublished master dissertation). Tehran: Imam Sadeq University.

Deldam, E. (1984). *Khalije Fars (Persian Gulf)*. Tehran: Novin.

Ebrahimi, M. (2016). *The British role in Iranian domestic politics (1951–1953)*. Cham: Springer International Publishing.

Eghtedari, A. (1991). *Joghrafiyay-e Tarikhi-ye Masqat va Oman (Historical geography of Masqat and Oman)*. Tehran: Donyaye ketab.

Elahi, H. (1991). *Khalij-e Fars va Masael-e an (Persian Gulf and its problems)* (Vol. 2). Tehran: Ghumes.

Eqbal Ashtiyani, A. (2005). *Tarikh-e Iran az Sadr-e Islam ta Enqeraz-e Qajariyye* (Iran's History from Early Islam to the Collapse of the Qajar Dynasty) (Tehran: Samt Publications).

Farahmand, A. (2007). Kargardani-ye Engelis dar Tajziyye Bahrain az Iran (The British role in the separation of Bahrain from Iran). *Journal of Tarikh-e Ravabet-e Khareji, 31,* 85–145.
Farid, N. (1978). *Asr-e Este'mar Zodaee (De-Colonization Era).* Tehran: Amir Kabir.
Ghaem Maghami, J. (1982). *Bahrain va Masael-e Khalig-e Fars (Bahrain and Persian Gulf problems).* Tehran: Tahouri.
Hedayat, M. G. (2006). *Khaterat (Memories.* Tehran: Zavvar.
Houshang Mahdavi, A. (2009). *Siyasat-e Khareji-ye Iran Dar Asr-e Pahlavi (Iran's foreign policy in Pahlavi Era).* Tehran: Alborz.
Ibn Battuta, M. (1997). *Safarnameh (Travel literature).* Tehran: Agah.
Kernoker, G. (2011). Jodaiy-e Bahrain az Iran, Naf'e Iran ya Engelis (The separation of Bahrain from Iran: Iran or British interests). *Faslname-ye Motale'at-e Tarikhi, 8*(33), 224.
Khosravi, A., & Alimardani, A. A. (2011). Este'mar-e Engelis va Jodaiy-e Bahrain az Iran (British Colonialism and Bahrain's separation from Iran). *Faslname-ye Motale'at-e Tarikhi, 15,* 167–193.
Kia, I. (2006). *Seyr-e Este'mar-e Engelis dar Iran (Trends of British Colonization in Iran).* Tehran: Abi.
Modak, M. R. (1966). *The lands and people of India* (F. Gorgani, Trans.). Tehran: Tarjemeh va Nashr-e Ketab.
Moghtader, H. (2007). Cheshm Poushi-ye Iran az Bahrain (Iran's renunciation of Bahrain). *Ettelaat-e Siyasi Eghtesadi, 22*(34).
Mojtahed-Zadeh, P. (2001). *Security and territoriality in the Persian Gulf: A maritime political geography.* Tehran: Foreign Ministry Institute for Political and International Studies.
Mojtahed-Zadeh, P. (2006). *Keshvarha Va Marzha dar Mantagheye Geopolitice Khalige Fars (Countries and borders in geopolitical region of Persian Gulf)* (Vol. 3). Tehran: Daftar-e Motale'at-e Vezarat-e Kharejeh.
Momeni, M., & Zamannejad, S. (2011). Barresiy-e Diplomaciy-e Engelis dar Jodaiy-e Bahrain az Iran (Analysis of British diplomacy role in separation of Bahrain from Iran). *Faslname-ye Motale'at-e Tarikhi, 8*(33), 209–222.
Monshi, E. (1998). *Tarikhe AlamAraye Abbasi (AlamAraye Abbasi history).* Tehran: Donyay-e Ketab.
Movahhed, M. A. (2001). *Mobaleghe-ye Este'mar (The Colonial Power Exaggeration).* Tehran: Karnameh.
Nehru, J. (1971). *Negahi be Tarikh-e Jahan (Glimpses of world history)* (M. Tafazoli, Trans.). Tehran: Amir Kabir.
Neshat, S. (1971). *Tarikh-e Siyasi-ye Khalig-e Fars (Persian Gulf political history).* Tehran: Kanoun-e Ketab.
Partovi Moghaddam, A. (2006). Chera va Chegune Bahrain az Iran Joda Shod (How and why Bahrain was separated from Iran). *Faslname-ye Motale'at-e Tarikhi, 15,* 81–111.
Ramazani, R. K. (1979). *The Persian Gulf and the strait of Hormuz.* Brill Archive.
Sadid Al Saltane, A. (1991). *Tarikh-e Masqat, Oman, Bahrain va Qatar va Ravabet-e Anha ba Iran (History of Masqat, Oman, Bahrain and Qatar and their relations with Iran).* Tehran: Eghtedari.
Salimi, A. (1922). *Asnad-e Mahramane-ye Siyasi-ye Engelis ya Siyasat-e Engelis va Rusiyye dar Iran Ghabl az Jang (British secret documents or British and Russia policy in Iran before the war).* Tehran: Bagherzade.
Tolouee, M. (1987). *Nabard-e ghodratha dar Khalij-e Fars (The great powers war in the Persian Gulf.* Tehran: Pyke Tarjeme va Nashr.
Valdani, A. J. (Ed.). (1991). *Barresi-ye Tarikhi-ye Ekhtelafat-e Iran va Iraq (Historical review of the Iran-Iraq Border dispute)* (Vol. 2). Tehran: Daftar-e Motale'at-e Vezarat-e Kharejeh.
Valdani, A. J. (1997). *Negahi Tarikhi be Jazayer-e Iran-ye Tonb va Abu Musa (A historical view of Iranian Tonb and Abu Musa).* Tehran: Daftar-e Motale'at-e Vezarat-e Kharejeh.
Vosoughi, M. B. (2007). *Tarikh-e Khalig-e Fars va Keshvarhay-e Hamjavar-e an (History of the Persian Gulf and its neighboring lands).* Tehran: Payam-e Nour.
Wilson, A. (1987). *Persian Gulf.* Tehran: Elmi Farhangi.
Yarshater, E. (2004). *Encyclopedia Iranica* (Vol. 3, p. 509).

Zargar, A. A. (1993). *Bahrain az Asr-e Baastan ta Emrouz (Bahrain from ancient times to the Modern Era)*. Tehran: Honarbakhsh.
Zarrin Qalam, A. (1958/1337). *Sarzamine Bahryn az Dorane Bastan ta Emrouz (Land of Bahrain from Encient time till Now)*. Tehran: Ketab-e Sabz.

Masoumeh Rad Goudarzi (Iran) is assistant professor at the Department of Political Science, Faculty of Literature and Humanities, University of Guilan (Iran). She specialises in Islamic Political Thought and Politics and Government in the Middle East and Caucasus. She holds a B.A. in political Science from Shahid Beheshti University, Tehran; an M.A. from Tehran Islamic Azad University; and a Ph.D. in Politics and Government from UPM, Malaysia. She is the author of *Ayatollah Khomeini and the Subject of Islamic Government* (Germany: Lambert Academic Publishing, 2011); has received several research awards, and is published in reputable peer-reviewed journals.

Sediqeh Nazarpour (Iran) lectures at Mazandarn University (Iran). She holds a B.A. in Political Science from Tehran Islamic Azad University, and an M.A. in Political Science from the Islamic Azad University of Chalous. Her research interests include identity studies, mass media, political sociology, and foreign policy.

Chapter 7
Conclusion: The Geopolitical Struggle for Iran, Central Asia and the Caspian Sea Basin

Saikou Kawsu Gassama, Mansoureh Ebrahimi and Kamaruzaman Yusoff

The authors have presented various issues related to Iran's geopolitical significance for the last four hundred-plus years. Iran has consistently attracted the interests of rival imperial powers, especially Russia and Britain, in an ongoing scramble for the dominion and control of its lands and resources. From the early nineteenth century through to 1970, Iran lost several territories due to this activity.

Tsarist Russia needed to secure its massive borders and also trade with neighbours in Central Asia, the Caspian Sea Basin, South Asia and Eurasian regions. It was determined to maintain these spheres of influence by any means necessary. A key British objective in Iran was to ensure permanent control of the country's Shahs and territories but without direct colonisation. Hence, it combined diplomatic means (treaties, deception, emissaries, and resident ambassadors) with coercive strategies (threats, aggression, temporal occupation and war) to protect British interests in India.

Britain failed to honour several treaties and refused to aid Iran during the Russo-Persian War that ended with the Treaty of Turkmenchay in 1828. This directly led to further losses of Persian territory in addition to reparation payments. Britain then signed a secret agreement with Russia in 1854 that split Iran into two spheres of influence. It also supported Iran's enemies during the war over Herat, and thus created Afghanistan. The threat of aggression ('imperial bullying') was common British currency whereby "the English Government could… make itself master of it [Kharak]

S. K. Gassama · M. Ebrahimi (✉) · K. Yusoff
Faculty of Islamic Civilization, Universiti Teknologi Malaysia, 81310 Johor Bahru, Johor, Malaysia
e-mail: mansoureh@utm.my; suriebrahimi@gmail.com
URL: https://www.linkedin.com/in/suri-ebrahimi-08aa0457/
URL: http://islamic.utm.my/mansoureh/
URL: http://afes-press-books.de/html/APESS_29.htm

S. K. Gassama
e-mail: skgassama2012@gmail.com

K. Yusoff
e-mail: y.kamaruzaman@utm.my; zuryati@yahoo.com

M. Ebrahimi et al. (eds.), *The Dynamics of Iranian Borders*,
https://doi.org/10.1007/978-3-319-89836-0_7

at any moment it chose, nor could all the world combined prevent its doing so, while that nation preserved superiority at Sea"[1] (Das 2016).

Britain saw Iran as a buffer against Russian expansionism from the reign of Peter The Great (1689–1725), a position which continued under Catherine the Great (1762–96) (Arndreeva 2007: 13–15), especially with regard to the Persian Gulf, Central Asia and the Middle East in general, as Russian influence posed deadly threats to British interests. Hence, London determined to maintain the status quo at all costs, including the use of militarism. This imperialist philosophy thus guided two rival forces on a collision course over Iran and Central Asia as an applied reality of Halford Mackinder's theory: The Heartland versus the World-Island: "Who rules East Europe commands the Heartland: Who rules the Heartland commands the World-Island: Who rules the World-Island commands the World (1919: 194)" (Cox et al. 2008: 7). As discussed in our Introduction, Mackinder's framework provides a succinct explanation for the continuum of Anglo-Russian relations and contentions for centuries, especially over Central Asia and Iranian borders.

Britain was equally apprehensive about Napoleon's intentions towards India and particularly alarmed by the peace treaty of Gulistan between France and Russia between Iran and Russia. Hence, core British policy was to pre-empt any French military expedition in Iran, Ottoman or Arabian regions in the Middle East. John Malcolm,[2] British Resident in Iran and chief negotiator, assessed Russia's means for invasion as 'great', and further stated that Iran's resources for assisting either a Russian or French invasion of India were '*very considerable*'. Hence, British plans were considerably dedicated to the security of its imperial interests in South Asia. As such, it used Persia as a buffer zone to secure British India's north-western frontier and protect trade and supply lines (Das 2016: 71). Malcolm effectively communicated alarm over the pending Franco-Persian alliance, especially since Britain had no opportunity to secure a tripartite alliance with Russia. Hence, Malcolm proposed the occupation of Khark Island to control the Persian Gulf as follows:

> Our establishment at Khark Island would fix us in that position in this quarter, which it appears to me as indispensable to our interests. We should assume and be enabled to act with a spirit of independence and prompt decision, which the times require, and thus be able to affect everything with the half-barbarians of this quarter. When they see us possessed of such effectual local means of either injury or support of their interests ... the jealousy excited by such a step would be considerable and would be exceeded by their fears. (Das 2016: 72)

[1]Malcolm delivered this message through Pasley and Dr. Andrew Jukes (surgeon and British Resident at Bushehr) to the Persians via Abdul Rasul, the Sheikh of Bushehr. Malcolm was prepared to forcefully occupy Kharat if he failed to do so by negotiations. Malcolm saw Kharat as the potential centre needed to secure British interests in the Persian Gulf. Although the Sheikh would welcome the British Flag, Malcolm became livid when the Sheikh refused him permission on the orders of the Shah of Persia.

[2]Minto proposed Col. John Malcolm for appointment as Britain's Political Agent to Persia and Turkish Arabia (30 January 1808). Minto was convinced that Malcolm could detach Persia from hostile enemies like Russia and France, who were bent on invading British India. Malcolm's mission was framed in such a way that it did not clash with Harford Jones, who was appointed as her Majesty's representative in Persia (Das 2016: 61).

It can be said that British imperial objectives in Eurasia, Iran and the Caspian Sea Basin, as well as in Central and South Asia and the Middle East, were mostly motivated by economics. Michael Mann's study in historical sociology, *The Sources of Social Power*, introduced the IEMP power model (Ideological, Economic, Military and Political) (Smith and Owens 2001: 277), which conveys a further appreciation of the British colonial context.

1. *Ideological Power*: derives from '*the human need to find ultimate meaning in life, to share norms and values.*' Hence, ideological power '*provides a transcendent vision of social authority*' (Mann 1986: 519) where control over ideology is the source of general social power.
2. *Economic Power*: the result of a need '*to extract, transform, distribute and consume the resources of nature*'.
3. *Military Power*: '*the social organisation of physical force*' that stems from the '*necessity of organised defence and the utility of aggression.*' Those who control or monopolise military power exercise some degree of general social power.
4. *Political Power*: state power that results from the '*usefulness of territorial and centralised regulation*' (Mann 1986: 277) with an authoritative central command.

The IEMP model postulates that, 'the struggle to control ideological, economic, military, and political power organisations provides the central drama of social development. Societies are structured primarily by entwined ideological, economic, military, and political power' (Mann 1986: 277). These four social power sources were ever-present features of the British imperial armoury and were unleashed in numerous ways, as clearly illustrated by the authors of this book. They were primarily applied to classical realist thinking, thus comprising a colonial policy decision-making machine that Martin Wight would call '… *the blood and iron and immorality men*' (Wight 1994).

The chief object of this heartless colonial policy in Iran and Central Asia was to secure and protect British India for trade, resource exploitation, taxation and the provision of labour in support of numerous colonial institutions and infrastructure. Hence, economic considerations were primary with other factors in formal attendance. Initially, Britain had to secure overland trade routes and supply lines from India across Iran and Central Asia; therefore, they secured India's north-west frontier. Britain then created sundry institutions and residencies throughout the entire region as a support network over which British control served to maintain its vital economic resources. Hobson observed the following:

> The total value of the import and export trade of Britain in 1898 (we take this year as the latest normal one for the purpose, later years being disturbed by war) amounted to £765,000,000. If we were to take the very liberal allowance of 5 per cent as profit upon this turnover of trade, the annual income directly derived from our external trade would amount to a little over £38,000,000, or about one forty-fifth part of our annual income. (Hobson 1902: 31)

British imperial efforts provided its islands' life-blood. These economic resources and attendant power enabled Britain to accumulate wealth and become the world's leading imperial power. Hobson noted: 'the absorption of so large a proportion of public interest, energy, blood and money in seeking to procure colonial possessions and foreign markets would seem to indicate that Britain obtains her livelihood by external trade' (Hobson 1902: 30).

Ideological power had been centred on centuries of conflict between Iran, Russia, Britain and France in which the role of religion (Islam vs. Christianity) proved critical. At different times, both enjoyed ascendancy and dominance and were often extremely tense as proven by the Crusade era.

Misinterpretation has been a hallmark of these relations since the emergence of Islam in the seventh century and the early encounters in the medieval period, when Islam was the dominant force. The balance of power has swung throughout the centuries—at times the Christian world had to respond to Muslim challenges in various forms, while at other times it was the Muslims who had to react to Christian initiatives. By the nineteenth century, the images and perceptions of each side towards the other were developing in the context of European colonialism and an increasingly unequal power relationship (Arndreeva 2007: 23).

European colonialists, to varying degrees, also believed they were on a civilising mission to other races. This world view dominated the '*survival of the fittest*' model in which social Darwinian comportment pretty much guided British colonial dealings with Iran and elsewhere. Indeed, it is no small wonder that Malcolm carried the white man's burden so well by referring to Iranians as 'barbarians' while serving in the foreign office.

While the British excelled at the 'indirect rule' of their colonies[3] by using 'Brown Sahibs' as 'Rajahs' in India and 'Chiefs' in Africa, the French were noted for their colonial policy of assimilation. Colonial administrations were thus all ideologically based and therefore exerted a form of social power over subjugated peoples. In Part II of his classic analysis of imperialism entitled *Imperialism: A Study* (1902),[4] Hobson

> investigates the theory and practice of imperialism regarded as a 'mission of civilisation', in its effect upon 'lower' or alien peoples, and its political and moral reactions upon the conduct and character of the Western nations engaging in it. (Hobson 1902: v)

It became apparent that in their struggles over Africa and Asia, imperial powers (Britain, France, Russia, Germany, Portugal, Holland, Italy and others) had amassed immense '*social power*' deriving from ideological propaganda in support of the so-called '*civilising*' mission.

This Eurocentric bias presumed a distinction between Western identity and Oriental 'others'. The world became dichotomised as us vs. them, civilization vs. barbarity, good vs. evil, reason vs. emotion, etc. The Orient was transformed into a passive object that silently and weakly submitted to civilised study and processes conducted by an advanced West. Russia also charged itself with the duty to 'civilise' and thus bring economic and political progress and Christianity to the peoples of Central

[3]For a detailed discussion of 'indirect rule', see (Lugard 1965).

[4]'Hobson was a historian and economic journalist who was one of the so-called 'New Liberals' who emerged at the end of the 19th century. The New Liberals advocated much greater state intervention in the economy than the classical liberals had done. One of the things that makes Hobson especially interesting is his support for Cobdenite ideas of peace and free trade' (Hobson 1902).

Asia and the Caucasus (Arndreeva 2007: 8–9). Obviously, '*civilizing*' required cruel subjugation through imperialism, a matter the late Professor Edward Said has thoroughly criticised in his monumental work, *Orientalism* (Said 1979).

The military dimensions of power were made all the more apparent in the scramble for control over Eurasia, Iran and Central Asia. Britain, Russia and France all used brute force at some stage in the pursuit of respective imperial agendas:

> The scramble for Africa and Asia has virtually re-cast the policy of all European nations, has evoked alliances that cross all natural lines of sympathy and historical association, has driven every continental nation to consume an ever-growing share of its material and human resources upon military and naval equipment, and has drawn the great new power of the United States from its isolation into the full tide of competition. (Hobson 1902: 11)

The US, though arriving late in the imperial melee, would become its biggest player by the end of WWII.

Military power's purpose is to aggressively support imperialism's ideological basis and policies, especially regarding economic plunder. Machiavellian political machinations and intrigues were therefore necessary for the realpolitik approach where 'ends justify the means'. Morality absent, this game's only rule was to win. Hence, rights or laws that respected victims were/are routinely ignored in favour of the sheer convenience and efficient pragmatism that obtains imperial goals. The authors have shown across various chapters that Russia fought several wars with Iran; that the British captured and occupied Khark Island; the ceding of Herat; the Crimean War (1853–56); and sundry other brute force measures that secured territorial aggrandisement in favour of imperial economic interests. Hobson further expounded on the classical realism of imperialism:

> While producing for popular consumption doctrines of national destiny and imperial missions of civilisation, contradictory in their true import, but subsidiary to one another as supports of popular Imperialism, it has evolved a calculating, greed type of Machiavellianism, entitled '*realpolitik*' in Germany, where it was made, which has remodelled the whole art of diplomacy and has erected national aggrandisement without pity or scruple as the conscious motivating force of foreign policy. Earth hunger and the scramble for markets are responsible for the openly avowed repudiation of treaty obligations, which Germany, Russia and Britain have not scrupled to defend. (Hobson 1902: 11)

Without military power, even if all the other sources of power were present, imperialism would have failed miserably. Hence, coercive subjugation by any means necessary actually crowned the 'civilising' efforts of colonial powers.

Politically, Britain, Russia, Portugal and France had been engaged in a diplomatic struggle, by any means, to attract and secure the Shah of Persia as 'friend and partner' via competitive treaties with favourable terms for respective interests. In the thick of this imperial chronicle, Iran was repeatedly victimised as a virtual colony over which they had all feuded for centuries. The authors have highlighted various treaties that served to sedulously erode Iran's time-honoured territorial domains. We noted also that British machinations often deployed deception and deliberate breaches of agreements that served to divide Central Asia and allow Britain's dominion. Consequent to such machinations was/is the lasting insecurity and '*loss of vast Iranian territories, some of which remain in unresolved dispute to this day*'.

Combining the four social sources of power of the IEMP model, imperialists imposed respective ideological roots in support of a wide array of economic networks and markets they subsequently sustained by realpolitik intrigues and brute force whenever necessary to maintain dominion over territorial aggrandisements in Iran as well as in Central and South Asia. While the British were fully aware of the strengths and weaknesses of their imperial power base in Iran during the Napoleonic period, Iranians failed to understand that these same weaknesses and fears could have saved them from British impositions and plunder. Iran often had weak leaders who effectively played into imperialist hands, no matter who called on the court. Britain was aware they were not a land power, which if fully understood by Iranians might have changed the power equation entirely. Hence, Malcolm generally favoured military aggression as opposed to Harford Jones,[5] who favoured diplomacy by treaty.

In the face of the French-Persian overture, Malcolm thought it necessary to leverage regional power in India for a partnership with Iran, especially since British power in continental Europe was weak. The Brigadier General found it most discomfiting that Iran had sent an Envoy to London on the advice of Jones, which would provide opportunity for Iran to assess Britain in military terms that could be disadvantageous vis-à-vis France and Russia (Hobson 1902: 86). But as all sides were pressuring the Shah, Malcolm knew Iran would soon realise the strategic significance of friendship with British India. In the absence of such an awakening, coercive measures would then be deployed:

> Should, however, the Shah of Persia be ever so far deluded as to give openly and decidedly a preference to the enemies of Britain [sic], that nation should, instead of having recourse to measure fruitless conciliation, which would undoubtedly be mistaken for proofs of weakness, take such steps as were calculated to awaken that Monarch to a just sense of the importance of its friendship. The establishment of a more intimate alliance with the Bashah [sic] of Baghdad, the withdrawing of a factory at Abushesher, the formation of a settlement at the Gulph (a point which is under every view of importance), and the temporary interruption of the intercourse between India and Persia, would either oblige the Shah of Persia to alter his policy, or throw his Dominions and life into great hazard. (Hobson 1902: 62)

In light of the foregoing discussion, we conclude that British policy in Iran and Central Asia was anchored in the policy to protect British India rather than any direct colonisation of Iran. The fear of losing the 'Jewel in the Crown' was aptly contextualised a century later after WWII by Field Marshal Viscount Archibald

---

[5]Canning, the British Foreign Secretary, appointed Sir Harwood Jones as His Majesty's Envoy to the Court of Persia in August 1807. Being familiar with Persian politics, Jones was designated 'Envoy-Extraordinary' and Minister-Plenipotentiary'. He had been an assistant to the Resident of Baghdad for nine years, had visited Shiraz, learnt Persian and also engaged in private trade alongside political office. Thus, he communicated with authorities on Persian affairs in Central Asia on the Russo-Persian War and during the beginning of French infiltration of Persia and Central Asia. His appointment was premised on a notion of the Court of Directors to the Foreign Office that royal credentials (Jones appointment as Royal Envoy) would carry greater weight in the Persian Royal Court as agent for both the Foreign Office and the East India Company. His term included counteracting French policies and determining the nature of their intrigues in Persia and threats to British India (Hobson 1902: 56–57).

Percival Wavell,[6] at a time when Britain was reluctant to grant independence to India in spite of growing popular demand. Wavell described the significance of India to the British Empire in a letter to Winston Churchill, October 1944:

> I feel very strongly that the future of India is the problem on which the British Commonwealth and the British reputation will stand or fall in the post-war period. To my mind our strategic security, our name in the world for statesmanship and fair dealing and much of our economic well-being will depend on the settlement we make in India. Our prestige and prospects in Burma, Malaya, China, and the Far East generally are entirely subject to what happens in India. If we can secure India as a friendly partner in the British Commonwealth, our predominant influence in these countries will, I think, be assured. But with a lost and hostile India, we are likely to be reduced in the East to the position of commercial bag-men. (Das 1982: 14)

This kind of strategic thinking informed and underpinned British imperial policy in Iran, Central Asia and South Asia. Hence, Britain was prepared to slug it out with Russia throughout its occupation of India in a two-way tussle for power and glory that perfectly fits Mackinder's theory regarding the battle for supremacy between the 'Heartland' and the 'World-Island'.

By combining might and classic realist diplomacy, Britain kept hold of its spheres of influence in Central Asia, Iran and Afghanistan with the primary objective of securing British India. It essentially Balkanised the Persian Empire with the Treaty of Paris (1857), which initiated the 'political separation of Iran and Afghanistan from their ancient Persian fealty'. By severing this centuries-old allegiance, Britain formally created Afghanistan for the first time in world history, as attested by official documents.

By the end of the nineteenth century there was no possibility for Iran to maintain its age-old suzerainty over its vast expanse. The dynamics of *realpolitik* had altered time-honoured status quos: 'The standard of justice depends on the equality of power to compel and that in fact the strong do what they have the power to do and the weak accept what they have to accept' (McGowan et al. 1999: 24). As the order of the day, Classical Realism had no regard for moral or legal restraints; hence, the Persian Empire gradually contracted. Even the Sykes-Picot (secret) Agreement (1916) between France and Britain placed Basra and Baghdad under Britain's direct control. Iran was forced to cede further territories and by 1970 Iran was reduced to its present size by the separation of Bahrain, also via British collusion.

However, we must point out that internal failings of local governors and leadership facilitated the process and loss of Persia's sovereignty and territories. The authors made this perfectly clear:

> [Re:] incompetence and ignorance of contemporary local governors and Iranian leadership… They were not only unaware of the actual geopolitical situation regionally and trans-regionally, but also lacked the ability to respond reasonably or impose more favourable treaty obligations. This inherently divisive association of Emirs, Shahs, Sultans and Governors also lacked knowledge regarding the legal principles of border partitioning. Hence, they forsook

[6] He became Viceroy and Governor General of India in 1943. Prior to this he was the Commander-in-Chief of British India as well as the India-Burma theatre Commander during WWII. See (Das 1982: 14; Bradford 2006: 278).

their responsibilities as leaders and lost what God had given them to protect for the want of knowledge and political unity, making them easy prey for the British Lion.

The British similarly employed deception and intrigue as imperial weapons in Africa (Boahen 1990: 99). Britain's unique geographical position, being detached from continental Europe, had spurred its pursuit of strategic '*commercial and political development overseas*' by dependence on a strong navy. 'Not only the power and the greatness, but the very safety and independence of Great Britain depend on her maritime strength' (Temperley and Penson 1966: 292).

As a result of the deadly rivalry between Britain and Russia, colonial expansionism by land and sea in the Caspian Sea Basin and Central Asia allowed for certain developments: 'As a result of the colonial expansion of Britain and Russia, by the late nineteenth century Iran had become a semi-colony. Tsarist Russia owned motor and railroads, banks, and many industrial enterprises in Iran' (Zonn et al. 2010: 225). On another note, although lamenting that 'geographers have allowed themselves to be pushed into political side lines and away from national and international affairs', mainly due to the underestimation of the profession's relevancy by geographers themselves, Slowe (1990) expressed the importance and relevance of geography and the environment as follows:

From World Bank to the parish council, from superpowers imposing their will on their neighbours to a provincial independence struggle in a famine-struck African state, it is geography that determines economic and political strategy and geography that determines cultural and military tactics. Only geography provides complete models of the world in which decisions are made. Only geographers focus on the relationships between culture, economy, society and polity, to model the whole landscape. (Slowe 1990: 1)

We might then surmise from the foregoing that geography and environment were of strategic significance to the achievement of political power[7] two centuries ago and that this relationship has not diminished in the contemporary scramble for hydrocarbon resources in the Caspian Sea. In Chaps. 4 and 5 the authors noted growing American governmental and corporate interests in sourcing alternative energy reserves removed from the volatile Middle East. US policy and the spreading of its sphere of influence into CIS republics, possibly including a NATO expansionist programme, has breached Russian sensitivities and alerted them to the trespass in their 'Backyard'. Thus, they are bent on using all necessary means to stop the intrusions.

The geographic presence of hydrocarbon energy is undergoing a radical shift that has initiated the struggle for Caspian Sea hydrocarbon resources; hence, perhaps another 'Great Game' has commenced. Joseph A. Stanislaw described the tenor of this game in which rules are still being written:

[7]Political geographers like Mackinder have discussed the notion of the state as the 'pivot' of political geography. 'The objective of geographic work in this tradition is to support and justify the extension of the power of the national state by outlining specific geographic features or areas that are crucial for political control' Herb (2008: 23). Nevertheless, 'The use of geography to project political power can be traced to Herodotus in the fourth century BCE and to Ibn Khaldun in the fourteenth century CE. Herodotus, who is considered the father of History, is also acclaimed as the father of geography.'

> Broad outlines are already clear. It is no longer confined to Central Asia, as was the original Great Game of the 19th century, pitting the British and Russian empires against each other. Today, while the energy resources of the Caspian feature prominently in the game, both the Chessboard and the number of players have expanded. (Stanislaw 2009: 3)

This suggests that no single country is in control of strategic power relations in the Caspian Basin, which therefore suggests that a time bomb ticks away with grave potential for an out-of-hand explosion if not effectively managed among competing neo-colonial forces. We no longer confront the limited resources of past imperial fancies. Times have changed and the potential for deadly rivalry and war are both imminent and ominous and reach far beyond the geographical limitations of modern-day Iran.

References

Arndreeva, E. (2007). *Russia and Iran in the Great Game: Travelogues and orientalism*. London: Routledge.

Boahen, A. (1990). *A general history of Africa VII: Africa under colonial domination 1880–1935*. London: James Currey Ltd/UNESCO, International Scientific Committee.

Bradford, J. C. (Ed.). (2006). *International encyclopedia of military history*. NY: Rutledge.

Cox, K. R., Low, M., Robinson, J. (2008). Introduction: Political geography: Traditions and turns. In K. R. Cox, M. Low, J. Robinson (Eds.), *The SAGE handbook of political geography*. London: SAGE Publications.

Das, M. N. (1982). *Partition and Independence of India: Inside story of the Mountbatten days*. New Delhi: Vision Books.

Das, A. (2016). *Defending British India against Napoleon: The foreign policy of Governor-General Lord Minto (1807–1813)*. Woodbridge: Boydell Press.

Herb, G. H. (2008). The politics and political geography. In K. R. Cox, M. Low, & J. Robinson (Eds.), *The SAGE handbook of political geography*. London: SAGE Publications.

Hobson, J. A. (1902). *Imperialism: A study*. NY: James Pott & Co.

Lugard, F. (1965). *The dual mandate in British Tropical Africa*. Abingdon: Frank Cass and Co.

Mann, M. (1986). *The sources of social power: Vol. 1. A history of power from the beginning to AD 1760*. Cambridge University Press.

McGowan, P. J., Cornelissen, S., & Nel, P. (Eds.). (1999). *Power, wealth and global equity: An international relations textbook for Africa*. Lansdowne: Institute for Global Dialogue, University of Captown Press.

Said, E. (1979). *Orientalism*. New York: Vintage Books.

Slowe, P. M. (1990). *Geography & political power: The geography of nations and states*. London: Routledge.

Smith, S., & Owens, P. (2001). Alternative approaches to international theory. In J. Baylis, S. Smith (Eds.), *The globalization of world politics*. Oxford: Oxford University Press.

Stanislaw, J. A. (2009). *Power play-resource nationalism, the global scramble for energy and the need for mutual interdependence*. Deloitte Development LLC: Deloitte Centre for Energy Solutions.

Temperley, H. W. V., & Penson, L. M. (1966). *Foundations of British Foreign Policy from Pitt (1792) to Salisbury (1902)*. London: Frank Cass & Co., Ltd.

Wight, M. (1994). *International theory: The three traditions*. Leicester: Leicester University Press.

Zonn, I. S., Kosarev, A. N., Glantz, M. H., & Kostianoy, A. G. (Eds.). (2010). *The Caspian Sea Encyclopedia*. Heidelberg: Springer.

Saikou Kawsu Gassama (Malaysia) is a Ph.D. student in the Faculty of Islamic Civilization, Universiti Teknologi Malaysia. He holds a B.A. in Political Science from the International Islamic University Malaysia, an M.A. in Islamic Political Economy from the University of Durham (2001), an M.A. in Social Science from the National University of Malaysia (2004), and an MA in Diplomacy, Law and Global Change from Coventry University (2007). He has worked as a Programme Specialist Governance and Partnership at Action Aid International in The Gambia (Sept 2013–Sept 2016), as a Lecturer at the Department of Political Science, School of Art and Sciences, The University of The Gambia, and served as the Secretary to the Cabinet, Office of the President of the Republic of the Gambia (Mar 2011–Mar 2012), and Permanent Secretary 1, at the Office of the President (Feb 2010–Mar 2011).

Mansoureh Ebrahimi (Iran/Malaysia) Ph.D., is a senior lecturer with the Faculty of Islamic Civilization, Universiti Teknologi Malaysia. Dr. Ebrahimi has written about Iran's culture, history, religious beliefs and Malaysia's *halal* industry from the Islamic perspective. She also teaches Persian and Philosophy of Science and Civilization. Her books include *The British Role in Iranian Domestic Politics (1951–1953)* (Germany: Springer Nature, 2016); Mansoureh Ebrahimi and Kamaruzaman Yusoff (Eds.): *The Halal Industry in Malaysia: Prospects and Challenges* (Malaysia: UTM Press, 2015); has also contributed several articles on culture, religion and history (including Malaysia, Middle East, Iran and Turkey) to reputable peer-reviewed journals; 'The British plot to remove Iran's democratically elected Prime Minister from power', in: *Acta Histriae* (Republic of Slovenia: Annales Publishing House), 23, (2015): 735–748.

Kamaruzaman Yusoff (Malaysia) is professor at the Faculty of Islamic Civilization, Universiti Teknologi Malaysia, Head of West Asia Civilization Research Group, where he specialises in Middle Eastern History and Islamic Studies. He holds a B.A. in History from the National University of Malaysia; and an M.Litt. and a D.Litt. from the Department of Islamic and Middle Eastern Studies, Edinburgh University, Scotland. He is the author of several articles and books on the Middle East, Orientalism in Islamic History, and the relationship between the Ottoman Empire and the Malay World; and has served as Head of Department Research Group and Dean of the Faculty.

Index

M. Ebrahimi et al. (eds.), *The Dynamics of Iranian Borders*,
https://doi.org/10.1007/978-3-319-89836-0

The manufacturer's authorised representative in the EU is Springer
Nature Customer Service Centre GmbH, Europaplatz 3, 69115 Heidelberg,
Germany. If you have any concerns regarding our products, please
contact ProductSafety@springernature.com

Printed and bound by CPI Group (UK) Ltd, Croydon, CR0 4YY

30/09/2026

02225664-0015